SYMMETRY & MODERN PHYSICS

Yang Retirement Symposium

SYMMETRY & MODERN PHYSICS

Yang Retirement Symposium

State University of New York, Stony Brook
21 – 22 May 1999

edited by

A. Goldhaber
R. Shrock
J. Smith
G. Sterman
P. van Nieuwenhuizen
W. Weisberger

SUNY, Stony Brook, USA

World Scientific
New Jersey • London • Singapore • Hong Kong

Published by

World Scientific Publishing Co. Pte. Ltd.
5 Toh Tuck Link, Singapore 596224
USA office: Suite 202, 1060 Main Street, River Edge, NJ 07661
UK office: 57 Shelton Street, Covent Garden, London WC2H 9HE

British Library Cataloguing-in-Publication Data
A catalogue record for this book is available from the British Library.

SYMMETRY AND MODERN PHYSICS
Yang Retirement Symposium

ISBN 981-238-503-7
ISBN 981-238-563-0 (pbk)

Printed in Singapore by Mainland Press

Professor Chen Ning Yang

FOREWORD

The career of Chen Ning Yang, extending from wartime China of the nineteen forties into this century, is one of the great stories of modern science. His arrival in Chicago in 1946, to study with Fermi, with Teller and other historic figures there was part of a new era of science, passing from Europe, through America, to the world. In a lifetime of research contributions, he has transformed the way in which scientists think of matter and energy. A pioneer in physics, his great achievements and international stature have served as an inspiration for generations of young scientists, especially from developing nations.

Chen Ning Yang's contributions to physics are unsurpassed in the latter half of the Twentieth Century in their scope and depth. The story of his analysis of parity nonconservation in the weak interactions with T. D. Lee is the stuff of scientific legend: their deciphering of then puzzling features of particle decay, the publication of their solution, and the thunderbolt of its confirmation leading to the Nobel Prize just a year later. This brought them to the world stage as the first Chinese to win that award.

Yang's development with R. L. Mills of the concept of non-abelian gauge invariance and of the class of Yang–Mills theories was another such landmark event. In this case, it took two decades for the concepts that they introduced to flower into the contemporary Standard Model of elementary particle physics. Even beyond the successes of the Standard Model, however, the work of Yang and Mills set physics on a road that it still travels today. They did this by introducing the concept that symmetry principles are a guide to as-yet undiscovered particles and forces. This instinct for the role of symmetry, for the identification of the most essential features, is also reflected in his many contributions to statistical mechanics crowned perhaps by the development of what is now known as the Yang–Baxter equation. That discovery first unravelled the structure of a basic many-body problem, but has echoed again and again in physics and mathematics over the succeeding years.

Frank (the English name he chose in admiration of Benjamin Franklin) served as Einstein Professor and Director of the Institute for Theoretical Physics at Stony Brook University for thirty-two years, from 1966 until 1998, by far his longest-held professional position. As a Nobel Laureate at the height of his creative powers he boldly accepted a position at a then new and nearly unknown university, thereby instantly bringing Stony Brook international attention. As Director, he created and guided a new institute. He set a research style that continues here today, and helped to launch numerous careers of the junior faculty, postdocs and students who passed through the Institute, each of whom came away with memories of encounters with one of the leaders of Twentieth Century science.

When Frank Yang announced his intention to retire from these positions, it was clear that the closing of this era would provide a unique opportunity to celebrate his unexcelled contributions to physics. It also quickly became clear that representing those fields of physics in which he had worked or that had been influenced by his work, would result in a symposium that transcended the boundaries of any specialization and indeed the separations between theoretical and experimental physics and between physics and mathematics. At the same time by tapping the admiration that Frank Yang has commanded and the reservoirs of visionary encouragement he has sown worldwide, this same effort resulted in the gathering of a superlative list of participants and speakers. Each of these distinguished careers had intersected with Frank's, some speakers having encountered him as a brilliant graduate student, some as the renowned statesman of science that he remains today.

The Symposium "Symmetries and Reflections" was held at Stony Brook on May 21–22, 1999. The list of participants speaks for itself. For those of us involved in the organization, there was a special reward in the participation of so many returning colleagues, those who had passed through C. N. Yang's Institute as the first, or nearly first, step on their personal paths in research. By returning at this season they helped celebrate not only Frank's career but their own as well, and the collaborative venture that is science. In retrospect, the presence of Robert L. Mills (1927–1999) who chaired the opening session and who engaged in an inspiring discussion with students during his visit, was particularly meaningful.

The collected articles that follow represent the breadth and depth of the Symposium. The reader will find both the excitement of discovery and the pleasure of recollection. There are numerous anecdotes and fond memories, some strongly-held opinions in physics and beyond, but most of all the kinds of concepts and questions that make science an adventure. The major themes of contemporary physics, and the major discoveries of recent decades are well represented. Some of these articles contain material that has been updated since the Symposium. We believe that this collection can serve as a snapshot of the frontiers of physics at the close of the Twentieth and the opening of the Twenty-first Century. We are delighted as well to present a complete list of Chen Ning Yang's publications up to the end of 2002.

As with any significant event involving so much travel, and so removed from the normal routine, the Symposium could not have happened without a dedicated group, sharing a common goal and vision. It was also made possible by a generous and skilled support staff, including Ms. Elizabeth Gasparino and Ms. Doreen Matesich. The organizers are grateful for the support of Stony Brook University, including its Provost Robert L. McGrath and President Shirley Kenny and of the Department of Physics and Astronomy including past Chairs Peter Kahn, Peter Paul and Janos Kirz. This volume would not have been possible without the energetic efforts of K.K. Phua of World Scientific Publishing, and we greatly appreciate this chance to preserve the record of the Symposium. We also thank Ms. Judy Wong of Hong Kong University for the list of the publications of C. N. Yang including titles in Chinese. Finally, we thank our colleagues, both present and former, from what is now, since the time of this Symposium, the C. N. Yang Institute for Theoretical Physics of Stony Brook University, for their support and for continuing the tradition set by Frank.

A. Goldhaber, R. Shrock, J. Smith, G. Sterman, P. van Nieuwenhuizen
& W. Weisberger
Stony Brook, USA
June, 2003

CONTENTS

Freeman Dyson

A CONSERVATIVE REVOLUTIONARY*

FREEMAN DYSON
Institute for Advanced Study, Princeton, NJ 08540, USA

Received 22 May 1999

I am delighted to have this opportunity to sing the praises of my old friend and colleague Frank Yang. The title of my talk is "A Conservative Revolutionary". The meaning of the title will become clear at the end of the talk.

One of my favorite books is Frank's "Selected Papers 1945–1980 with Commentary", published in 1983 to celebrate his sixtieth birthday. This is an anthology of Frank's writings, with a commentary written by him to explain the circumstances in which they were written. There was room in the book for only one third of his writings. He chose which papers to include, and his choices give a far truer picture of his mind and character than one would derive from a collection chosen by a committee of experts. Some of the chosen papers are important and others are unimportant. Some are technical and others are popular. Every one of them is a gem. Frank was not trying to cram as much hard science as possible into five hundred pages. He was trying to show us in five hundred pages the spirit of a great scientist, and he magnificently succeeded. The papers that he chose show us his personal struggles as well as his scientific achievements. They show us the deep sources of his achievements, his pride in the Chinese culture that raised him, his reverence for his teachers in China and in America, his love of formal mathematical beauty, his ability to bridge the gap between the mundane world of experimental physics and the abstract world of groups and fiber bundles. He wisely placed the eighty pages of commentaries together at the beginning of the book instead of attaching them to the individual papers. As a result, the commentaries can be read consecutively. They give us the story of Frank's life in the form of an intellectual autobiography. The autobiography is a classic. It describes the facts of his life in clear and simple words. It quietly reveals the intense feelings and loyalties that inspired his work and made him what he is.

One of the smallest and brightest of the gems in Frank's book is a two-page description of Fermi, written as an introduction to a paper by Fermi and Yang that

*Talk at the banquet of the C. N. Yang retirement symposium, May 21–22, 1999, Stony Brook.

was included in a volume of Fermi's collected papers. Frank studied with Fermi in Chicago from 1946 to 1949. He learned more physics from Fermi than from anybody else, and Fermi's way of thinking left an indelible impression in his mind. Frank writes, "We learned that physics should not be a specialist's subject. Physics is to be built from the ground up, brick by brick, layer by layer. We learned that abstractions come after detailed foundation work, not before". Fermi's practical spirit can be seen in the title of the great Yang–Mills paper published in 1954. Anyone speaking about the paper today would call it the paper that introduced non-Abelian gauge fields. But the title does not mention non-Abelian gauge fields. The title is "Conservation of isotopic spin and isotopic gauge invariance". The physical question, how to understand the conservation of isotopic spin, came first, and the mathematical abstraction, non-Abelian gauge fields, came second. That was the way Fermi would have approached the problem, and that was the way Frank approached it too. Fermi was great because he knew how to do calculations and also knew how to listen to what nature had to say. All through his life, Frank has balanced his own gift for mathematical abstraction with Fermi's down-to-earth attention to physical details.

Please let me digress here briefly to tell a story about Fermi that has nothing to do with Frank. I was not a student of Fermi, but I had the good luck to spend 20 minutes with Fermi at a crucial point in my career. I learned more from Fermi in 20 minutes than I learned from Oppenheimer in 20 years. In 1952 I thought I had a good theory of strong interactions. I had organized an army of Cornell students and post-docs to do calculations of meson–proton scattering with the new theory. Our calculations agreed pretty well with the cross-sections that Fermi was then measuring with the Chicago cyclotron. So I proudly traveled from Ithaca to Chicago to show him our results. Fermi was polite and friendly but was not impressed. He said, "There are two ways to do calculations. The first way, which I prefer, is to have a clear physical picture. The second way is to have a rigorous mathematical formalism. You have neither". That was the end of the conversation and of our theory. It turned out later that our theory could not have been right because it took no account of vector interactions. Fermi saw intuitively that it had to be wrong. In 20 minutes, his common sense saved us from several years of fruitless calculations. This was a lesson that Frank did not need to learn, since he had already absorbed Fermi's common sense during his years as a student in Chicago.

Frank has not been idle during the 15 years since his selected papers were published. Another book was published in 1995, this time not written by Frank but by his friends, a festschrift to celebrate his seventieth birthday, with the title "Chen Ning Yang, a great physicist of the twentieth century". This book contains, hidden among the technical contributions, a number of personal tributes and recollections. It describes Frank's active involvement, continuing up to the present day, helping science to grow and flourish in three Chinese communities, in the People's Republic of China, in Taiwan and in Hong Kong. Frank is happy to be able to pay back the debt that he owes to his native land and culture.

Not included in either of the two books is a paper written by Frank two years ago with the title "My father and I". This is a tribute to his father, who was a professor of mathematics and died in 1973. It is a wonderfully sensitive account of his relationship to his father and of the pain that each of them suffered as a result of their separation. His father stayed in China through the hard years while Frank grew to greatness in America. Both of them knew that it was better so. Without America, Frank could not have become a world-class scientist. Exiled from China, his father would have been a tree without roots. And yet, the separation hurt both of them deeply. For Frank, his personal separation from his father and the political separation of America from China were two parts of a single tragedy. Luckily, President Nixon decided to recognize the People's Republic just in time, so that Frank was able to visit China twice before his father died and to sit by his bedside during his last illness. In the commentary to his Selected Papers, Frank describes the difficult decision that he made in 1964 to become a citizen of the United States. This was a formal recognition of his separation from China and from his father. He writes, "My father ... had earned a Ph.D. degree from the University of Chicago in 1928. He was well traveled. Yet I know, in one corner of his heart, he did not forgive me to his dying day for having renounced my country of birth".

The memoir "My father and I" ends on a happier note. It ends with a glorious moment of reunion. Frank describes how he stood at midnight on July 1, 1997, at the Hong Kong Convention and Exhibition Center, to watch the Union Jack being lowered and the flag of the People's Republic being slowly raised, while the band played "Arise, you who would not be enslaved". Frank writes, "Had father observed this historical ceremony marking the renaissance of the Chinese people, he would have been even more moved than I. ... The intellectuals of his generation had to personally experience the humiliating exploitations in the Foreign Concessions ... and countless other rampant foreign oppressions. ... How they had looked forward to the day when a prosperous China could stand up, when the British Empire had to lower the Union Jack and withdraw troops, when they can see for themselves the Chinese flag proudly announce to the world: This is Chinese Territory! That day, July the first, 1997, is the day their generation had dreamed of throughout their lives".

We can all rejoice that Frank was standing there to give his blessing and his father's blessing to the reunion. For me, that pride and that feeling of fulfilment that Frank expresses have a special resonance. I too belong to a great and ancient civilization. My home-town in England was also the home-town of Alfred the scholar king, who made our town into a great center of learning eleven hundred years ago, while the Tang dynasty was establishing the system of government by scholars that endured for a thousand years in China. Our king Alfred was translating scholarly texts from Latin into English, soon after the Tang poet Tu Fu wrote the poem that Frank quotes at the beginning of his Selected Papers: "A piece of literature is meant for the millennium. But its ups and downs are known already in the author's heart".

Like Frank, I too left my homeland and became an American citizen. I still remember the humiliation of that day in Trenton when I took the oath of allegiance to the United States, and the ignoramus who performed the ceremony congratulated me for having escaped from the land of slavery to the land of freedom. With great difficult I restrained myself from shouting out loud that my ancestors freed our slaves long before his ancestors freed theirs. I share Frank's ambivalent feelings toward the United States, this country that has treated us both with so much generosity and has treated our ancient civilizations with so little understanding. And I share Frank's pride in the peaceful lowering of the Union Jack and raising of the Chinese flag that he witnessed in Hong Kong, the place where our two ancient civilizations briefly came together and gave birth to something new.

Five years ago, I had the honor of speaking at the ceremony in Philadelphia, when the Franklin Medal was awarded to Frank Yang by the American Philosophical Society. We were assembled in the historic meeting-room of the society, with the portraits of Benjamin Franklin, the founder of the society, and Thomas Jefferson, one of its most active members, looking down at us. It was self-evident that Franklin and Jefferson approved of the award. We know that Frank Yang feels a special admiration for Franklin, since he gave the name of Franklin to his elder son. I would like to end this little talk with the same words that I used to praise Frank on that happy occasion.

Professor Yang is, after Einstein and Dirac, the preeminent stylist of the 20th century physics. From his early days as a student in China to his later years as the sage of Stony Brook, he has always been guided in his thinking by a love of exact analysis and formal mathematical beauty. This love led him to his most profound and original contribution to physics, the discovery with Robert Mills of non-Abelian gauge fields. With the passage of time, his discovery of non-Abelian gauge fields is gradually emerging as a greater and more important event than the spectacular discovery of parity non-conservation which earned him the Nobel Prize. The discovery of parity non-conservation, the discovery that left-handed and right-handed gloves do not behave in all respects symmetrically, was a brilliant act of demolition, a breaking-down of intellectual barriers that had stood in the way of progress. In contrast, the discovery of non-Abelian gauge fields was a laying of foundations for new intellectual structures that have taken 30 years to build. The nature of matter as described in modern theories and confirmed by modern experiments is a soup of non-Abelian gauge fields, held together by the mathematical symmetries that Yang first conjectured 45 years ago.

In science, as in urban renewal and international politics, it is easier to demolish old structures than to build enduring new ones. Revolutionary leaders may be divided into two kinds, those like Robespierre and Lenin who demolished more than they built, and those like Benjamin Franklin and George Washington who built more than they demolished. There is no doubt that Yang belongs to the second kind of revolutionary. He is a conservative revolutionary. Like his fellow-revolutionaries Franklin and Washington, he cherished the past and demolishes as little as possible. He cherishes with equal reverence the great intellectual traditions of Western science and the great cultural traditions of his ancestors in China.

Yang likes to quote the words of Einstein, "The creative principle lies in mathematics. In a certain sense, therefore, I hold it true that pure thought can grasp reality, as the ancients dreamed". On another occasion Yang said, "That taste and style have so much to do with one's contribution in physics may sound strange at first, since physics is supposed to deal objectively with the physical universe. But the physical universe has structure, and one's perceptions of this structure, one's partiality to some of its characteristics and aversion to others, are precisely the elements that make up one's taste. Thus it is not surprising that taste and style are so important in scientific research, as they are in literature, art and music". Yang's taste for mathematical beauty shines through all his work. It turns his least important calculations into miniature works of art, and turns his deeper speculations into masterpieces. It enables him, as it enabled Einstein and Dirac, to see a little further than other people into the mysterious workings of nature.

J. Zinn-Justin

RENORMALIZATION OF GAUGE THEORIES AND MASTER EQUATION

J. ZINN-JUSTIN

DAPNIA, CEA/Saclay, F-91191, Gif-sur-Yvette Cedex, France*

and

Institut de Mathématiques de Jussieu–Chevaleret,
Université de Paris VII, France
zinn@spht.saclay.cea.fr

The evolution of ideas which has led from the first proofs of the renormalizability of non-Abelian gauge theories, based on Slavnov–Taylor identities, to the modern proof based on the BRS symmetry and the *master equation* is briefly recalled. The content and consequences of the master equation are explained. This lecture has been delivered at the Symposium in Honor of Professor C. N. Yang, Stony Brook, May 21–22, 1999.

1. Introduction

It is a rare privilege for me to open this conference in honor of Professor Yang. His scientific contributions have been for me an essential source of inspiration. The most obvious example, Yang–Mills fields or gauge theories, will be illustrated by my talk. But there are other important aspects of his work which have also directly influenced me: Professor Yang has consistently shown us that a theorist could contribute to quite different domains of physics like particle physics, the statistical physics of phase transitions or integrable systems .... Moreover his work has always emphasized mathematical elegance.

Finally by offering me a position at the ITP in Stony Brook in 1971, he has given me the opportunity to start with the late Benjamin W. Lee a work on the renormalization of gauge theories, which has kept me busy for several years and played a major role in my scientific career.

Let me add a few other personal words. The academic year 1971–1972 I spent here at the ITP has been one of the most exciting and memorable of my scientific life. One reason of course is my successful collaboration with Ben Lee. However, another reason is the specially stimulating atmosphere Professor Yang had managed to create at the ITP, by attracting talented physicists, both ITP members and visitors, by the style of scientific discussions, seminars and lectures.

My interest in Yang–Mills fields actually dates back to 1969, and in 1970 I started a work, very much in the spirit of the original paper of Yang and Mills,

*Laboratoire de la Direction des Sciences de la Matière du Commissariat à l'Energie Atomique, France.

on the application of massive Yang–Mills fields to strong interaction dynamics. Although in our work massive Yang–Mills fields were treated in the spirit of effective field theories, we were aware of the fact that such quantum field theories were not renormalizable.

In the summer of 1970 I presented the preliminary results of our work in a summer school in Cargèse, where Ben Lee was lecturing on the renormalization of spontaneous and linear symmetry breaking. This had the consequence that one year later I arrived here at the ITP to work with him.

Ben had just learned, in a conference I believe, from 't Hooft's latest work on the renormalizability of non-Abelian gauge theories both in the symmetric and spontaneously broken phase and was busy proving renormalizability of the Abelian Higgs model. We immediately started our work on the much more involved non-Abelian extension.

Our work was based on functional integrals and other powerful functional methods, in contrast to less reliable and much less transparent manipulations of Feynman diagrams, and a generalization of so-called Slavnov–Taylor identities, consequence of the properties of the Faddeev–Popov (FP) determinant arising in the quantization of gauge theories. In a series of four papers (1972–1973), we examined most aspects of the renormalization of gauge theories.

Notation. In this lecture, we will always use an euclidean formalism, and thus will not distinguish betweeen space and time.

2. Classical Gauge Action and Quantization

The principle of gauge invariance, which promotes a global (or rigid) symmetry under a Lie group G to a local (gauge) symmetry, provides a beautiful geometric method to generate interactions between relativistic quantum particles.

The basic field is a Yang–Mills or gauge field (mathematically a connection) $\mathbf{A}_\mu(x)$, related to infinitesimal parallel transport, and written here as a matrix belonging to the Lie algebra of the symmetry group

$$\mathbf{A}_\mu(x) = \sum_\alpha A_\mu^\alpha(x)\mathbf{t}_\alpha \,, \tag{1}$$

where the matrices $\mathbf{t}_\alpha$ are the generators of the Lie algebra of G in some representation.

Acting on the gauge field, a gauge transformation characterized by space-dependent group element $\mathbf{g}(x)$, takes an affine form:

$$\mathbf{A}_\mu(x) \mapsto \mathbf{g}(x)\mathbf{A}_\mu(x)\mathbf{g}^{-1}(x) + \mathbf{g}(x)\partial_\mu\mathbf{g}^{-1}(x) \,. \tag{2}$$

In particular, from the point of view of global transformations ($\mathbf{g}(x)$ constant), the field $\mathbf{A}_\mu(x)$ transforms by the adjoint representation of the group G.

To gauge transformations are associated *covariant* derivatives, whose form depends on the group representation, for example,

$$\mathbf{D}_\mu = \partial_\mu + \mathbf{A}_\mu \,.$$

They transform linearly under a gauge transformation:

$$\mathbf{D}_\mu \mapsto \mathbf{g}(x)\mathbf{D}_\mu\,\mathbf{g}^{-1}(x)\,. \tag{3}$$

The curvature $\mathbf{F}_{\mu\nu}(x)$ associated to the gauge field can be obtained from the covariant derivative by

$$\mathbf{F}_{\mu\nu}(x) = [\mathbf{D}_\mu, \mathbf{D}_\nu] = \partial_\mu \mathbf{A}_\nu - \partial_\nu \mathbf{A}_\mu + [\mathbf{A}_\mu, \mathbf{A}_\nu]\ .$$

It thus transforms linearly as

$$\mathbf{F}_{\mu\nu}(x) \mapsto \mathbf{g}(x)\mathbf{F}_{\mu\nu}(x)\,\mathbf{g}^{-1}(x)\,.$$

The pure Yang–Mills action is the simplest gauge invariant action. It can be written as

$$\mathcal{S}(\mathbf{A}_\mu) = -\frac{1}{4e^2}\int d^d x\,\mathrm{tr}\mathbf{F}^2_{\mu\nu}(x)\,,$$

where e is here the gauge coupling constant.

Matter fields that transform nontrivially under the group are coupled to the gauge field because gauge invariance dictates that derivatives must be replaced by covariant derivatives. For fermions the action takes the typical form

$$\mathcal{S}_{\mathrm{F}}(\bar\psi,\psi) = -\int d^d x\,\bar\psi(x)\,(\not{\mathbf{D}} + M)\,\psi(x)\,,$$

and for the boson fields:

$$\mathcal{S}_{\mathrm{B}}(\phi) = \int d^d x\left[(\mathbf{D}_\mu\phi)^\dagger\,\mathbf{D}_\mu\phi + V(\phi)\right],$$

in which $V(\phi)$ is a group invariant function of the scalar field ϕ.

Quantization. The classical action results from a beautiful construction, but the quantization apparently completely destroys the geometric structure. Due to the gauge invariance, the degrees of freedom associated with gauge transformations have no dynamics and, therefore, a straightforward quantization of the classical action does not generate a meaningful perturbation theory (though nonperturbative calculations in lattice regularized gauge theories can be performed). It is thus necessary to *fix* the gauge, a way of expressing that some dynamics has to be provided for these degrees of freedom. For example, motivated by quantum electrodynamics, one may add to the action a covariant nongauge invariant contribution

$$\mathcal{S}_{\mathrm{gauge}} = \frac{1}{2\xi e^2}\int d^d x\,\mathrm{tr}\,(\partial_\mu \mathbf{A}_\mu)^2\ . \tag{4}$$

However, simultaneously, and this is a specificity of non-Abelian gauge theories, it is necessary to modify the functional integration measure of the gauge field to maintain formal unitarity. In the case of the gauge (4), one finds

$$[d\mathbf{A}_\mu(x)] \mapsto [d\mathbf{A}_\mu(x)] \det \mathbf{M}\,, \tag{5}$$

where $\mathbf{M}$ is the operator

$$\mathbf{M}(x,y) = \partial_\mu \mathbf{D}_\mu \delta(x-y)\,.$$

This (Faddeev–Popov) determinant is the source of many difficulties. Indeed, after quantizing the theory one has to renormalize it. Renormalization is a theory of deformations of local actions. However, the determinant generates a nonlocal contribution to the action. Of course, using a well-known trick, it is possible to rewrite the determinant as resulting from the integration over un-physical spin-less fermions $\mathbf{C}, \bar{\mathbf{C}}$ (the "ghosts") of an additional contribution to the action

$$\mathcal{S}_{\text{ghosts}} = \int d^d x\, \bar{\mathbf{C}}(x) \partial_\mu \mathbf{D}_\mu \mathbf{C}(x)\,.$$

After this transformation the action is local and renormalizable in the sense of power counting. However, in this local form all traces of the original symmetry seem to have been lost.

3. Renormalization

The measure (5) is the invariant measure for a set of nonlocal transformations, which for infinitesimal transformations takes the form

$$\delta \mathbf{A}_\mu(x) = \int dy\, \mathbf{D}_\mu \mathbf{M}^{-1}(x,y)\omega(y),$$

the field $\omega(x)$ parametrizing the transformation. Using this property it is possible to derive a set of Ward–Takahashi (Slavnov–Taylor) identities between Green's functions and to prove renormalizability of gauge theories both in the symmetric and spontaneously broken Higgs phase. The nonlocal character of these transformations and the necessity of using two different representations, one nonlocal but with invariance properties, the other one local and thus suitable for power counting analysis, explains the complexity of the initial proofs.

Though the problem of renormalizing gauge theories could then be considered as solved, one of the remaining problems was that the proofs, even in the most synthetic presentation like in Lee–Zinn-Justin IV, were quite complicated, and more based on trial and error than systematic methods.

Returning to Saclay, I tried to systematize the renormalization program of quantum field theories with symmetries. I abandoned the idea of a determination of renormalization constants by relations between Green's functions, for a more systematic approach based on loop expansion and counter-terms.

The idea is to proceed by induction on the number of loops. Quickly summarized:

One starts from a regularized local Lagrangian with some symmetry properties. One derives, as consequence of the symmetry, identities (generally called Ward–Takahashi (WT) identities) satisfied by the generating functional Γ of proper vertices or one-particle irreducible (1PI) Green's functions. By letting the cut-off go to infinity (or the dimension to 4 in dimensional regularization), one obtains identities satisfied by the sum Γ_{div} of all divergent contributions at one-loop order. At this order Γ_{div} is a local functional of a degree determined by power counting. By subtracting Γ_{div} from the action, one obtains a theory finite at one-loop order. One then reads off the symmetry of the Lagrangian renormalized at one-loop order and repeats the procedure to renormalize at two-loop order. The renormalization program is then based on determining general identities, valid both for the action and the 1PI functional, which are stable under renormalization, that is stable under all deformations allowed by power counting. One finally proves the stability by induction on the number of loops.

Unfortunately, this program did not apply in an obvious way to non-Abelian gauge theories, because it required a symmetry of the local quantized action, and none was apparent. WT identities were established using symmetry properties of the theory in the nonlocal representation.

In the spring of 1974, my student Zuber drew my attention to a preliminary report of a work by Becchi, Rouet and Stora who had discovered a strange fermion-type (like supersymmetry) symmetry of the complete quantized action including the ghost contributions. There were indications that this symmetry could be used to somewhat simplify the algebra of the proof of renormalization. Some time later, facing the daunting prospect of lecturing about renormalization of gauge theories and explaining the proofs to nonexperts, I decided to study the BRS symmetry. I then realized that the BRS symmetry was the key allowing the application of the general renormalization scheme and in a summer school in Bonn (1974) I presented a general proof of renormalizability of gauge theories based on BRS symmetry and the *master equation*.

4. BRS Symmetry

The form of the BRS transformations in the case of non-Abelian gauge transformations is rather involved and hides their simple origin. Thus, we first give here a presentation which shows how BRS symmetry arises in apparently a simpler context.

4.1. *The origin of BRS symmetry: constraint equations*

Let φ^α be a set of dynamical variables satisfying a system of equations

$$E_\alpha(\varphi) = 0\,, \tag{6}$$

where the functions $E_\alpha(\varphi)$ are smooth, and $E_\alpha = E_\alpha(\varphi)$ is a one-to-one map in some neighborhood of $E_\alpha = 0$ which can be inverted into $\varphi^\alpha = \varphi^\alpha(E)$. This implies, in particular, that Eq. (6) has a unique solution $\varphi_{\rm s}^{\alpha}$. We then consider some function $F(\varphi)$ and we look for a formal representation of $F(\varphi_{\rm s})$, which does not require solving Eq. (6) explicitly.

One formal expression is

$$F(\varphi_{\rm s}) = \int \Bigl\{\prod_\alpha dE^\alpha\, \delta(E_\alpha)\Bigr\} F(\varphi(E)) = \int \Bigl\{\prod_\alpha d\varphi^\alpha\, \delta\left[E_\alpha(\varphi)\right]\Bigr\} \mathcal{J}(\varphi)\, F(\varphi) \tag{7}$$

with

$$\mathcal{J}(\varphi) = \det \mathbf{E}\,, \quad E_{\alpha\beta} \equiv \frac{\partial E_\alpha}{\partial \varphi^\beta}.$$

We have chosen $E_\alpha(\varphi)$ such that $\det \mathbf{E}$ is positive.

4.2. *Slavnov–Taylor identity*

The measure

$$d\rho(\varphi) = \mathcal{J}(\varphi) \prod_\alpha d\varphi^\alpha\,, \tag{8}$$

has a simple property. The measure $\prod_\alpha dE_\alpha$ is the invariant measure for the group of translations $E_\alpha \mapsto E_\alpha + \nu_\alpha$. It follows that $d\rho(\varphi)$ is the invariant measure for the translation group nonlinearly realized on the new coordinates φ_α (provided ν_α is small enough):

$$\varphi^\alpha \mapsto \varphi'^\alpha \quad \text{with} \quad E_\alpha(\varphi') - \nu_\alpha = E_\alpha(\varphi)\,. \tag{9}$$

This property is, in gauge theories, the origin of the Slavnov–Taylor symmetry.

The infinitesimal form of the transformation can be written more explicitly as

$$\delta\varphi^\alpha = [E^{-1}(\varphi)]^{\alpha\beta}\nu_\beta\,. \tag{10}$$

4.3. *BRS symmetry*

Let us again start from identity (7) and first replace the δ-function by its Fourier representation:

$$\prod_\alpha \delta\left[E_\alpha(\varphi)\right] = \int \prod_\alpha \frac{d\lambda^\alpha}{2i\pi} e^{-\lambda^\alpha E_\alpha(\varphi)}\,. \tag{11}$$

The λ-integration runs along the imaginary axis. From the rules of fermion integration, we know that we can also write the determinant as an integral over Grassmann variables c^α and $\bar{c}^\alpha$:

$$\det \mathbf{E} = \int \prod_\alpha (dc^\alpha d\bar{c}^\alpha) \exp\left(\bar{c}^\alpha E_{\alpha\beta} c^\beta\right)\,. \tag{12}$$

Expression (7) then takes the apparently more complicated form

$$F(\varphi_{\rm s}) = \mathcal{N} \int \prod_{\alpha} (d\varphi^{\alpha} dc^{\alpha} d\bar{c}^{\alpha} d\lambda^{\alpha})\, F(\varphi) \exp\left[-S(\varphi, c, \bar{c}, \lambda)\right] , \tag{13}$$

in which $\mathcal{N}$ is a constant normalization factor and $S(\varphi, c, \bar{c}, \lambda)$ the quantity

$$S(\varphi, c, \bar{c}, \lambda) = \lambda^{\alpha} E_{\alpha}(\varphi) - \bar{c}^{\alpha} E_{\alpha\beta}(\varphi) c^{\beta} . \tag{14}$$

While we seem to have replaced a simple problem by a more complicated one, in fact, in many situations (and this includes the case where Eq. (6) is a field equation) it is easy to work with the integral representation (13).

Quite surprisingly, the function S has a symmetry, which actually is a consequence of the invariance of the measure (8) under the group of transformations (10). This BRS symmetry, first discovered in the quantization of gauge theories by Becchi, Rouet and Stora (BRS), is a fermionic symmetry in the sense that it transforms commuting variables into Grassmann variables and vice versa. The parameter of the transformation is a Grassmann variable, an anti-commuting constant $\bar{\varepsilon}$. The variations of the various dynamic variables are

$$\begin{cases} \delta\varphi^{\alpha} = \bar{\varepsilon} c^{\alpha} , & \delta c^{\alpha} = 0 , \\ \delta\bar{c}^{\alpha} = \bar{\varepsilon}\lambda^{\alpha} , & \delta\lambda^{\alpha} = 0 \end{cases} \tag{15}$$

with

$$\bar{\varepsilon}^2 = 0 , \qquad \bar{\varepsilon} c^{\alpha} + c^{\alpha}\bar{\varepsilon} = 0 , \qquad \bar{\varepsilon}\bar{c}^{\alpha} + \bar{\varepsilon}\bar{c}^{\alpha} = 0 .$$

The transformation is obviously *nilpotent* of vanishing square: $\delta^2 = 0$.

When acting on functions of $\{\varphi, c, \bar{c}, \lambda\}$, the BRS transformation can be represented by a Grassmann differential operator

$$\mathcal{D} = c^{\alpha} \frac{\partial}{\partial\varphi^{\alpha}} + \lambda^{\alpha} \frac{\partial}{\partial\bar{c}^{\alpha}} . \tag{16}$$

The operator $\mathcal{D}$ has the form of a cohomology operator since

$$\mathcal{D}^2 = 0 , \tag{17}$$

and is the source of a BRS cohomology.

4.4. *BRS symmetry in gauge theories*

For simplicity, we consider form now on only pure gauge theories, but the generalization in presence of matter fields is simple.

In gauge theories, the role of the φ variables is played by the group elements which parametrize gauge transformations and Eq. (6) is simply the gauge fixing equation. In the example of the gauge choice (4), it reads

$$\partial_{\mu}\mathbf{A}_{\mu}(x) = \nu(x),$$

where ν is a stochastic field with gaussian measure.

The corresponding quantized action can be written as

$$\mathcal{S}(\mathbf{A}_\mu, \bar{\mathbf{C}}, \mathbf{C}, \lambda) = \int d^4x \, \mathrm{tr}\left[-\frac{1}{4e^2}\mathbf{F}^2_{\mu\nu} + \frac{\xi e^2}{2}\lambda^2(x) + \lambda(x)\partial_\mu \mathbf{A}_\mu(x) + \mathbf{C}(x)\partial_\mu \mathbf{D}_\mu \bar{\mathbf{C}}(x) \right]. \tag{18}$$

The form of BRS transformations that leave the action invariant,

$$\begin{cases} \delta \mathbf{A}_\mu(x) = -\bar{\varepsilon}\mathbf{D}_\mu \mathbf{C}(x)\,, & \delta \mathbf{C}(x) = \bar{\varepsilon}\mathbf{C}^2(x), \\ \delta \bar{\mathbf{C}}(x) = \bar{\varepsilon}\lambda(x), & \delta\lambda(x) = 0\,, \end{cases} \tag{19}$$

is more complicated only because they are expressed in terms of group elements instead of coordinates.

Introducing the BRS differential operator

$$\mathcal{D} = \int d^dx \, \mathrm{tr}\left[-\mathbf{D}_\mu \mathbf{C}(x) \frac{\delta}{\delta \mathbf{A}_\mu(x)} + \mathbf{C}^2(x) \frac{\delta}{\delta \mathbf{C}(x)} + \lambda(x) \frac{\delta}{\delta \bar{\mathbf{C}}(x)} \right], \tag{20}$$

one can also express the BRS symmetry of the quantized action by the equation

$$\mathcal{D}\mathcal{S}(\mathbf{A}_\mu, \bar{\mathbf{C}}, \mathbf{C}, \lambda) = 0\,. \tag{21}$$

Moreover, one shows quite generally that the nongauge contrbution $\mathcal{S}_{\text{gauge}}$, which results from quantization, is BRS exact. Here,

$$\mathcal{S}_{\text{gauge}} = \mathcal{D} \int d^dx \, \mathrm{tr}\, \bar{\mathbf{C}}(x) \left[\partial_\mu \mathbf{A}_\mu(x) + \xi e^2 \lambda(x) \right]. \tag{22}$$

5. Renormalization and Master Equation

A set of WT identities that can be used to prove the renormalizability of gauge theories now correspond to BRS symmetry.

WT identities are based on change of variables of the form of BRS transformations. As a consequence, they involve the composite operators (nonlinear local functions of the fields) $\mathbf{D}_\mu \mathbf{C}(x)$ and $\mathbf{C}^2(x)$, which appear in the r.h.s. of the BRS transformation (19) and which necessitate additionnal renormalizations. To discuss renormalization, it is thus necessary to add to the action two sources $\mathbf{K}_\mu$, $\mathbf{L}$ for them:

$$\mathcal{S}(\mathbf{A}_\mu, \mathbf{C}, \bar{\mathbf{C}}, \lambda) \mapsto \mathcal{S}(\mathbf{A}_\mu, \mathbf{C}, \bar{\mathbf{C}}, \lambda) + \int d^4x \, \mathrm{tr}\left(-\mathbf{K}_\mu(x)\mathbf{D}_\mu \mathbf{C}(x) + \mathbf{L}(x)\mathbf{C}^2(x) \right).$$

The sources for BRS transformations, $\mathbf{K}_\mu$ and $\mathbf{L}$, have been later renamed anti-fields. No other terms are required because the two composite operators are BRS invariant.

5.1. *Master equation*

Since the composite operators $\mathbf{D}_\mu \mathbf{C}(x)$ and $\mathbf{C}^2(x)$ require renormalizations, the form (19) of BRS transformations is not stable under renormalization.

Instead, one discovers that the complete action $\mathcal{S}(\mathbf{A}_\mu, \mathbf{C}, \bar{\mathbf{C}}, \lambda, \mathbf{K}_\mu, \mathbf{L})$, which includes these additional source terms, satisfies after renormalization a quadratic relation, the master equation, which does not involve the explicit form of the BRS transformations (19). In component form (generalizing the notation (1) to all fields), the master equation reads

$$\int d^4x \sum_\alpha \left(\frac{\delta \mathcal{S}}{\delta A^\alpha_\mu(x)} \frac{\delta \mathcal{S}}{\delta K^\alpha_\mu(x)} + \frac{\delta \mathcal{S}}{\delta C^\alpha(x)} \frac{\delta \mathcal{S}}{\delta L^\alpha(x)} + \lambda^\alpha(x) \frac{\delta \mathcal{S}}{\delta \bar{C}^\alpha(x)} \right) = 0 . \qquad (23)$$

The proof involves first showing that this equation implies a similar equation for the generating functional of proper vertices (the 1PI functional). The latter equation, in turn, implies relations between divergences. In the framework of the loop expansion, these relations imply that the counter-terms that have to be added to the action to cancel divergences, can be chosen such that the renormalized action still preserves the master equation.

What is striking is that the master Eq. (23) contains no explicit reference to the initial gauge transformations. Therefore, one might worry that it does not determine the renormalized action completely, and that the general renormalization program fails in the case of non-Abelian gauge theories. However, one slowly discovers that the master equation has remarkable properties. In particular, all its local solutions which satisfy the power counting requirements and ghost number conservation, have indeed the form of an action for a quantized non-Abelian gauge theory.

Power counting relies on the canonical dimensions of fields, which are (with the notation $[X]$ for the canonical dimension of X)

$$[\mathbf{A}] = 1 , \quad [\mathbf{C} + \bar{\mathbf{C}}] = 2 , \quad [\lambda] = 2 , \quad [\mathbf{K}_\mu + \mathbf{C}] = 3 , \quad [\mathbf{L} + \mathbf{C} + \bar{\mathbf{C}}] = 4 ,$$

where this form takes into account ghost number conservation by displaying only the dimensions of the relevant products.

The action density has dimension 4 and, therefore, is quadratic in λ and linear in $\mathbf{K}_\mu$ and $\mathbf{L}$. The master equation implies that the coefficients $f_{\alpha\beta\gamma}$ of $L^\alpha C^\beta \bar{C}^\gamma$ are constants that satisfy the Jacobi identity, and the coefficients $D^{\alpha\beta}_\mu$ of $K^\alpha_\mu C^\beta$ are affine functions of $\mathbf{A}_\mu$ that satisfy Lie algebra commutations with $f_{\alpha\beta\gamma}$ as structure constants. Then continuity implies, in the semi-simple example at least, preservation of all geometric properties.

One somewhat surprising outcome of analysis is that the master equation has, for general gauge fixing functions of dimension 2, solutions with quartic ghost interactions, which cannot be obviously related to a determinant. On the other hand the master equation (and this is one of its main properties) implies directly that the nongauge invariant part of the quantized action is BRS exact (like in Eq. (22)). This property then ensures gauge independence and unitarity.

Only a few years later, elaborating on a remark of Slavnov, was I able to reproduce a general quartic ghost term as resulting from a generalized gauge fixing procedure (Zinn-Justin, 1984).

After the renormalization program was successfully completed, one important problem remained, of relevance for instance to the description of deep-inelastic scattering experiments: the renormalization of gauge invariant operators of dimension higher than 4. Using similar techniques Stern–Kluberg and Zuber were able to solve the problem for operators of dimension 6 and conjecture the general form. Only recently has the general conjecture been proven rigourously by nontrivial cohomology techniques (Barnich, Brandt and Henneaux, 1995).

Acknowledgments

To Professor C. N. Yang as a testimony of admiration and gratitude, Stony Brook, May 21, 1999.

Bibliographical Notes

A preliminary account of this lecture has been published in

J. Zinn-Justin, *Mod. Phys. Lett.* **A19** (1999) 1227 [hep-th/9906115].

A number of relevant articles and reviews can be found below.

After the fundamental article

C. N. Yang and R. L. Mills, *Phys. Rev.* **96** (1954) 191,

the main issue was the quantization of gauge theories

R. P. Feynman, *Acta Phys. Polon.* **24** (1963) 697;
B. S. DeWitt, *Phys. Rev.* **162** (1967) 1195, 1239;
L. D. Faddeev and V. N. Popov, *Phys. Lett.* **B25** (1967) 29;
S. Mandelstam, *Phys. Rev.* **175** (1968) 1580.

In the following article we tried to apply the idea of massive Yang–Mills fields to Strong Interaction Dynamics

J. L. Basdevant and J. Zinn-Justin, *Phys. Rev.* **D3** (1971) 1865.

Among the articles discussing Ward–Takahashi and renormalization see for instance

G. 't Hooft, *Nucl. Phys.* **B33** (1971) 173; *ibid.* **B35** (1971) 167.
A. A. Slavnov, *Theor. Math. Phys.* **10** (1972) 99.
J. C. Taylor, *Nucl. Phys.* **B33** (1971) 436.
B. W. Lee and J. Zinn-Justin, *Phys. Rev.* **D5** (1972) 3121, *ibid.* 3137, *ibid.* 3155; *ibid.* **D7** (1973) 1049.
G. 't Hooft and M. Veltman, *Nucl. Phys.* **B50** (1972) 318.
D. A. Ross and J. C. Taylor, *Nucl. Phys.* **B51** (1973) 125;
B. W. Lee, *Phys. Lett.* **B46** (1973) 214; *Phys. Rev.* **D9** (1974) 933.

The anti-commuting type symmetry of the quantized action is exhibited in

C. Becchi, A. Rouet and R. Stora, *Comm. Math. Phys.* **42** (1975) 127.

Most of the preceding articles are reprinted in

Selected papers on *Gauge Theory of Weak and Electromagnetic Interactions*, ed. C. H. Lai (World Scientific, Singapore, 1981).

The general proof, based on BRS symmetry and the master equation, of renormalizability in an arbitrary gauge, can be found in the proceedings of the Bonn summer school 1974,

J. Zinn-Justin in *Trends in Elementary Particle Physics, Lecture Notes in Physics* **37**, eds. H. Rollnik and K. Dietz (Springer-Verlag, Berlin, 1975).

See also

J. Zinn-Justin in *Proc. 12th School of Theoretical Physics, Karpacz 1975*, Acta Universitatis Wratislaviensis 368;

B. W. Lee in *Methods in Field Theory*, Les Houches 1975, eds. R. Balian and J. Zinn-Justin (North-Holland, Amsterdam, 1976).

Finally a systematic presentation can be found in

J. Zinn-Justin, *Quantum Field Theory and Critical Phenomena*, Clarendon Press (Oxford 1989, 4th edn., 2002).

For an an alternative proof based on BRS symmetry and the BPHZ formalism see

C. Becchi, A. Rouet and R. Stora, *Ann. Phys. (NY)* **98** (1976) 287.

Nonlinear gauges and the origin of quartic ghost terms are investigated in

J. Zinn-Justin, *Nucl. Phys.* **B246** (1984) 246.

Renormalization of gauge invariant operators and the BRST cohomology are discussed in

H. Kluberg-Stern and J.-B. Zuber, *Phys. Rev.* **D12** (1975) 467;
G. Barnich, and M. Henneaux, *Phys. Rev. Lett.* **72** (1994) 1588;
G. Barnich, F. Brandt and M. Henneaux, *Comm. Math. Phys.* **174** (1995) 93.

Shou-Cheng Zhang

HIGH T_c SUPERCONDUCTIVITY: SYMMETRIES AND REFLECTIONS

SHOU-CHENG ZHANG

Department of Physics, McCullough Bldg., Stanford University, Stanford, CA 94305-4045, USA

Received 7 November 1999

This is a talk given at the Symposium "Symmetries and Reflections", dedicated to Prof. C. N. Yang's retirement. In this talk, I shall reflect on my personal interaction with Prof. Yang since my graduate career at SUNY Stony Brook, and his profound impact on my understanding of theoretical physics. I shall also review the SO(5) theory of high T_c superconductivity and show how my collaboration with Prof. Yang in 1990 led to the foundation of this idea.

It is a great honor for me to speak at the Symposium "Symmetries and Reflections", dedicated to Prof. C. N. Yang's retirement. In front of an audience of luminaries whose work shaped the physics of this half century, it is hardly fit to present my own work. Instead, I would like to take this opportunity to reflect on my personal encounter with Prof. Yang and his great impact on my career. Ever since my first exposure to physics as a school child, he has been a great role model for me. This beautiful spring day on Long Island reminds me vividly of my first day arriving at Stony Brook as a graduate student. Ever since that day, he taught me physics in classrooms and in private conversations. He taught me through his generous advice and inspiring examples. But most important of all, he taught me that physics is beautiful, and his great career shows to all of us that subjective judgement of beauty, taste and style can often lead to great discoveries in theoretical physics.

When I started school, China was still in the dark shadow of the culture revolution. However, even though we could not learn much about science at school, every school child knows the names of Yang and Lee, and their great contribution to science. I was so overwhelmed by the fact that scientist can "prove" nature's "left-handedness", and immediately decided to spend all my after school time to learn about physics. In those dark days of China's past, I found refuge in the beauty of science. I am forever grateful that Prof. Yang's influence in China made it possible.

Only one semester after entering China's Fudan University, I had the rare opportunity of a undergraduate scholarship to study at the Freie Universität Berlin. After finishing my undergraduate degree, both Prof. Meng, the only ethnic Chinese professor at FU Berlin and a long time friend of Prof. Yang, and Prof. Schrader, my thesis advisor, urged me to continue my graduate education at Stony Brook. It was a dream coming true when I received the admission papers from Stony Brook.

Even though he was extremely busy, Prof. Yang always made himself available to talk to the incoming graduate students. In my first meeting with him, he asked me what my interest was. I answered that the highest goal of theoretical physics is to pursue Einstein's dream and unify gravity with all other forces, and this is what I am interested in doing. To my great shock, not only did he discourage me from doing that, he even discouraged me from doing particle physics in general. He argued eloquently that physics is a broad discipline, and interesting problems can be found everywhere.

I left Prof. Yang's office in a deep crisis and confusion. Fortunately, he offered a course called "Selected Problems in Theoretical Physics" to the incoming graduate students. In this course, he did not talk at all about what I consider to be the "frontier problems" in physics, but instead he talked about problems like Bohm–Aharonov effect, duality of the Ising model, flux quantization of a superconductor, phases and holography, off-diagonal-long-range-order and the concept of gauge field and magnetic monopoles. What I learned the most in this course was the selection of the topics. These topics reflected his personal taste in physics, and that's something one cannot easily learn from books. In these topics I saw that the complexity of nature can be captured by simplicity and beauty of the theory. This is what theoretical physics is all about.

I started working with Prof. van Nieuwenhuizen on supergravity, and quickly finished a few papers. However, partly because of Prof. Yang's advice, and partly because of the infectious enthusiasm of Prof. Steve Kivelson, I got more and more interested in condensed matter physics. I then became a postdoc at ITP UCSB. Prof. Schrieffer was kind enough to welcome a total newcomer into his group, and generously taught me about condensed matter physics. I had great fun working with him and Xiaogang Wen on the spin bag theory of high T_c, and finally decided to completely switch to condensed matter physics.

After UCSB, I joined the IBM Almaden Research Lab. At that time, Prof. Yang's paper[1] on "η pairing" just appeared in PRL, and I read it with great interest. I was fascinated by the mathematical simplicity and impressed by the fact that one can construct some exact eigenstates of an interacting Hamiltonian. I also noticed that since the η operator is a exact eigen-operator of the Hubbard model, $\eta^\dagger\eta$ should commute with the Hamiltonian, implying a new conservation law. But I had no idea what this means physically. I invited Prof. Yang to give a colloquium at IBM and discussed with him about my observation. He showed great interest in this observation, but we had to cut short our discussion since he had to catch a plane.

The next day I received a 15 page fax from Prof. Yang. On his plane back to New York, he worked out the complete mathematical theory of the SO(4) symmetry of the Hubbard model. It turns out that the η operator, its Hermitian conjugate and the total number operator form the generator of a "pseudospin" SU(2) algebra,

$$[\eta_0, \eta^\dagger] = \eta^\dagger \qquad [\eta_0, \eta] = -\eta \qquad [\eta^\dagger, \eta] = 2\eta_0 , \tag{1}$$

where $\eta_0 = (N_e - N)/2$ is the total number of electrons measured from half-filling. My observation that $\eta^\dagger \eta$ commutes with the Hamiltonian is nothing but the statement that the Casmir operator of this "pseudospin" SU(2) algebra is a conserved quantity. The ordinary SU(2) spin algebra and this pseudospin SU(2) algebra together form the complete SO(4) = SU(2) $\times$ SU(2) symmetry of the Hubbard model. I felt deeply honored and fortunate when he asked me to co-author the paper on the "SO(4) symmetry in the Hubbard model".[2]

What is the physical meaning of this pseudospin SU(2) symmetry? After our SO(4) paper, I spent a lot of time thinking about this question. This brings us to a seminal paper by Prof. Yang, on the "Concept of off-diagonal-long-range order and the quantum phase of liquid He and of superconductors".[3] In this seminal paper, he classified all forms of order in quantum many-body systems in terms of diagonal-long-range order (DLRO), like the spin-density-wave and charge-density-wave order, and off-diagonal-long-range order (ODLRO), like the superfluid and superconducting order. The DLRO is characterized by an order parameter with zero charge, but ODLRO in fermion systems is characterized by an order parameter which carries two units of charge. Almost all cooperative phenomena in condensed matter systems fall into either one of these two categories. These two different forms of order have very different physical manifestations and have fundamentally different mathematical characterizations. They are thought to be as different from each other as "night and day". Thercfore it came as a great surprise to me when I realized that the pseudo-spin symmetry exactly unifies these two different forms of order. Since the η operator carries charge two, it can connect DLRO and OLDLRO, whose charge quantum number differs exactly by two. To be more precise, there exists a relation between the η operator, the charge-density-wave order parameter Δ_c and the superconducting order parameter Δ_s,

$$[\eta^\dagger, \Delta_c] = i\Delta_s \, . \tag{2}$$

Therefore, the pseudospin SU(2) algebra performs a "rotation" between these two fundamentally different forms of order. Furthermore, this equation enables me to make a direct physical prediction. Since the ordinary U(1) charge symmetry is enlarged to a pseudospin SU(2) symmetry, the usual phase Goldstone mode has a symmetry partner, which I called the "η" collective mode. It exists in momentum space near (π, π) and is infinitely sharp in energy. It shows up only below the superconducting transition temperature T_c.[4]

While I was greatly excited by these observations, the condensed matter community was not. These observations were viewed as pure mathematical artifacts without realistic values. I was rather discouraged and did not continue this line of thinking. However, at the moments of solitude, I would quietly open Prof. Yang's book of "Selected Papers". I admired his style of physics, the beauty of symmetry and the power of mathematical reasoning. In looking at the examples from his work, I was deeply convinced that elegant mathematical concepts will eventually find their applications in physical systems.

In 1995, Prof. Doug Scalapino came to Stanford and reported on a new experimental discovery in high T_c superconductors. Using neutron scattering, experimentalists discovered a sharp, resolution limited resonance peak, which is centered around momentum (π, π) and energy of 40 meV. Furthermore, it appears only below the superconducting transition temperature.[5] I was greatly excited about this discovery, since this key signature of the resonance peak almost exactly matches my original prediction based on the pseudospin symmetry argument. Together with my student Eugene Demler, we quickly worked out a generalization of the pseudo-Goldstone mode theory, which offers a natural explanation of the neutron resonance peak in the high T_c superconductors.[6] The key difference between this work and my previous work on the η mode is the spin quantum number of the collective mode. While the η mode is a total spin singlet, this new mode is a spin triplet. In order to be consistent with Prof. Yang's notation, we later called this new collective mode the π mode. (In particle physics, the π mesons form a isospin triplet, while the η meson is a isospin singlet).

After this work with Eugene, I began to ask the following question: If the η mode is a consequence of the SO(4) symmetry, which symmetry dictates the π mode? This is the question which eventually leads to the formulation of the SO(5) theory of the high T_c superconductivity. I started commuting the π operators among themselves, but their commutators look rather complicated. Since I forgot most of my group theory, I could not recognize whether they form a simple Lie algebra or not. Few days later I was glancing at the phase diagram of high T_c, with the antiferromagnetic (AF) and the superconducting (SC) phases right next to each other, I suddenly realized that the new symmetry is supposed to unify DLRO and ODLRO. Since AF order parameter has 3 real components and the SC order parameter has 2 real components, the natural symmetry group should be SO(5)! After this realization, the organization of the commutators into a SO(5) Lie algebra became a trivial matter, and it really works!

Based on this simple concept, I began to formulate a theory of high T_c superconductivity using SO(5) as an effective symmetry unifying AF and SC order.[7] It not only unifies these two seemingly different forms of order and explains their close proximity in the high T_c systems, but also makes a number of striking experimental predictions. Currently, this theory is being tested both numerically and experimentally, and has the potential to ultimately solve the deep mystery of the high T_c superconductivity. Prof. Yang once summarized the fundamental laws in physics deduced in this century by a simple slogan "symmetry dictates interactions". If the SO(5) theory proves to be successful in condensed matter physics, it may be a example of "symmetry dictates phase diagrams". Needless to say, having such a higher organization principle could definitely guide our search and organization of diverse materials and their phases.

I would like to use this opportunity to thank Prof. Yang for his profound influence on my career and wish him happy retirement. His inspiration and his advice will always be my guide in the search for beauty and truth in physics.

References

1. C. N. Yang, *Phys. Rev. Lett.* **63**, 2144 (1989).
2. C. N. Yang and S. C. Zhang, *Mod. Phys. Lett.* **B4**, 759 (1990).
3. C. N. Yang, *Rev. Mod. Phys.* **34**, 339 (1962).
4. S. C. Zhang, *Phys. Rev. Lett.* **65**, 120 (1990); S. C. Zhang, *Int. J. Mod. Phys.* **B5**, 153 (1991).
5. J. Rossat-Mignod, L. Regnault, C. Vettier, P. Bourges, P. Burlet, J. Bossy, J. Henry and G. Lapertot, *Physica* **C185–189**, 86 (1991); H. Mook, M. Yethiraj, G. Aeppli and T. Mason, *Phys. Rev. Lett.* **70**, 3490 (1993); H. F. Fong, B. Keimer, P. W. Anderson, D. Reznik, F. Dogan and I. A. Aksay, *ibid.* **75**, 316 (1995); P. Dai, M. Yethiraj, F. A. Mook, T. B. Lindemer and F. Dogan, *ibid.* **77**, 5425 (1996); H. F. Fong, B. Keimer, D. L. Milius and I. A. Aksay, *ibid.* **78**, 713 (1997).
6. E. Demler and S.-C. Zhang, *Phys. Rev. Lett.* **76**, 4126 (1995).
7. S. C. Zhang, *Science* **275**, 1089 (1997).

Wolfgang Ketterle

WHEN ATOMS BEHAVE AS WAVES: BOSE–EINSTEIN CONDENSATION AND THE ATOM LASER*

WOLFGANG KETTERLE†

Department of Physics, MIT-Harvard Center for Ultracold Atoms,
and
Research Laboratory of Electronics, Massachusetts Institute of Technology,
Cambridge, Massachusetts, 02139, USA

Received 8 December 2001

1. Introduction

The lure of lower temperatures has attracted physicists for the past century, and with each advance towards absolute zero, new and rich physics has emerged. Laypeople may wonder why "freezing cold" is not cold enough. But imagine how many aspects of nature we would miss if we lived on the surface of the sun. Without inventing refrigerators, we would only know gaseous matter and never observe liquids or solids, and miss the beauty of snowflakes. Cooling to normal earthly temperatures reveals these dramatically different states of matter, but this is only the beginning: many more states appear with further cooling. The approach into the kelvin range was rewarded with the discovery of superconductivity in 1911 and of superfluidity in ^{4}He in 1938. Cooling into the millikelvin regime revealed the superfluidity of ^{3}He in 1972. The advent of laser cooling in the 1980's opened up a new approach to ultralow temperature physics. Microkelvin samples of dilute atom clouds were generated and used for precision measurements and studies of ultracold collisions. Nanokelvin temperatures were necessary to explore quantum-degenerate gases, such as Bose–Einstein condensates first realized in 1995. Each of these achievements in cooling has been a major advance, and recognized with a Nobel prize.

This paper describes the discovery and study of Bose–Einstein condensates (BEC) in atomic gases from my personal perspective. Since 1995, this field has grown explosively, drawing researchers from the communities of atomic physics, quantum optics, and condensed matter physics. The trapped ultracold vapor has emerged as a new quantum system that is unique in the precision and flexibility with which it can be controlled and manipulated. At least thirty groups have now

*Published with permission from the Nobel foundation, Sweden (2001).
†URL: http://cua.mit.edu/ketterle_group/

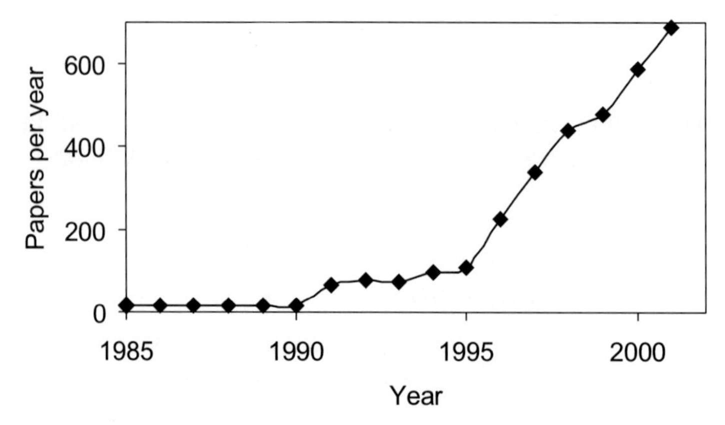

Fig. 1. Annual number of published papers, which have the words "Bose" and "Einstein" in their title, abstracts or keywords. The data were obtained by searching the ISI (Institute for Scientific Information) database.

created condensates, and the publication rate on Bose–Einstein condensation has soared following the discovery of the gaseous condensates in 1995 (see Fig. 1).

The phenomenon of Bose–Einstein condensation was predicted long ago, in a 1925 paper by Albert Einstein[1] using a method introduced by Satyendra Nath Bose to derive the black-body spectrum.[2] When a gas of bosonic atoms is cooled below a critical temperature T_c, a large fraction of the atoms condenses in the lowest quantum state. Atoms at temperature T and with mass m can be regarded as quantum-mechanical wavepackets that have a spatial extent on the order of a thermal de Broglie wavelength $\lambda_{\mathrm{dB}} = (2\pi\hbar^2/mk_{\mathrm{B}}T)^{1/2}$. The value of λ_{dB} is the position uncertainty associated with the thermal momentum distribution and increases with decreasing temperature. When atoms are cooled to the point where λ_{dB} is comparable to the interatomic separation, the atomic wavepackets "overlap" and the gas starts to become a "quantum soup" of indistinguishable particles. Bosonic atoms undergo a quantum-mechanical phase transition and form a Bose–Einstein condensate (Fig. 2), a cloud of atoms all occupying the same quantum mechanical state at a precise temperature (which, for an ideal gas, is related to the peak atomic density n by $n\lambda_{\mathrm{dB}}^3 = 2.612$). If the atoms are fermions, cooling gradually brings the gas closer to being a "Fermi sea" in which exactly one atom occupies each low-energy state.

Creating a BEC is thus simple in principle: make a gas extremely cold until the atomic wave packets start to overlap! However, in most cases quantum degeneracy would simply be pre-empted by the more familiar transitions to a liquid or solid. This more conventional condensation into a liquid and solid can only be avoided at extremely low densities, about a hundred-thousandth the density of normal air. Under those conditions, the formation time of molecules or clusters by three-body collisions (which is proportional to the inverse density squared) is stretched to

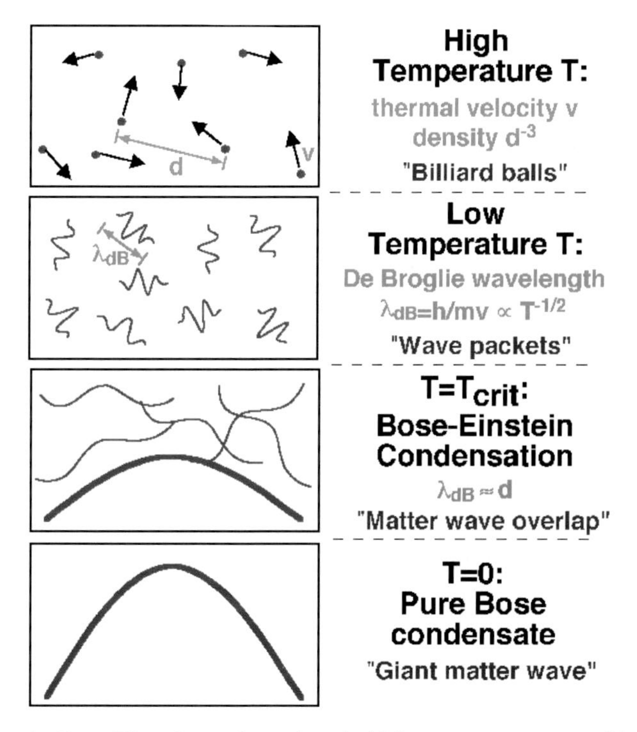

Fig. 2. Criterion for Bose–Einstein condensation. At high temperatures, a weakly interacting gas can be treated as a system of "billiard balls". In a simplified quantum description, the atoms can be regarded as wavepackets with an extension of their de Broglie wavelength λ_{dB}. At the BEC transition temperature, λ_{dB} becomes comparable to the distance between atoms, and a Bose condensate forms. As the temperature approaches zero, the thermal cloud disappears, leaving a pure Bose condensate.

seconds or minutes. Since the rate of binary elastic collisions drops only proportional to the density, these collisions are much more frequent. Therefore, thermal equilibrium of the translational degree of freedom of the atomic gas is reached much faster than chemical equilibrium, and quantum degeneracy can be achieved in an effectively metastable gas phase. However, such ultralow density lowers the temperature requirement for quantum degeneracy into the nano- to microkelvin range.

The achievement of Bose–Einstein condensation required first the identification of an atomic system which would stay gaseous all the way to the BEC transition, and second, the development of cooling and trapping techniques to reach the required regime of temperature and density. Even around 1990, it was not certain that nature would provide us with such a system. Indeed, many people doubted that BEC could ever be achieved, and it was regarded as an elusive goal. Many believed that pursuing BEC would result in new and interesting physics, but whenever one would come close, some new phenomenon or technical limitation would show up. A news article in 1994 quoted Steve Chu: "I am betting on nature to hide Bose condensation from us. The last 15 years she's been doing a great job".[3]

In brief, the conditions for BEC in alkali gases are reached by combining two cooling methods. Laser cooling is used to precool the gas. The principle of laser cooling is that scattered photons are on average blue-shifted with respect to the incident laser beam. As a result, the scattered light carries away more energy than has been absorbed by the atoms, resulting in net cooling. Blue-shifts are caused by Doppler shifts or AC Stark shifts. The different laser cooling schemes are described in the 1997 Nobel lectures in physics.[4–6] After the precooling, the atoms are cold enough to be confined in a magnetic trap. Wall-free confinement is necessary, otherwise the atoms would stick to the surface of the container. It is noteworthy that similar magnetic confinement is also used for plasmas which are too hot for any material container. After magnetically trapping the atoms, forced evaporative cooling is applied as the second cooling stage.[7–9] In this scheme, the trap depth is reduced, allowing the most energetic atoms to escape while the remainder rethermalize at steadily lower temperatures. Most BEC experiments reach quantum degeneracy between 500 nK and 2 μK, at densities between 10^{14} and 10^{15} cm^{-3}. The largest condensates are of 100 million atoms for sodium, and a billion for hydrogen; the smallest are just a few hundred atoms. Depending on the magnetic trap, the shape of the condensate is either approximately round, with a diameter of 10 to 50 μm, or cigar-shaped with about 15 μm in diameter and 300 μm in length. The full cooling cycle that produces a condensate may take from a few seconds to as long as several minutes.

After this short overview, I want to provide the historical context for the search for BEC and then describe the developments which led to the observation of BEC in sodium at MIT. Finally, some examples will illustrate the novel physics which has been explored using Bose–Einstein condensates. A more detailed account of the work of my group has been presented in four comprehensive review papers.[8,10–12]

2. BEC and Condensed Matter Physics

Bose–Einstein condensation is one of the most intriguing phenomena predicted by quantum statistical mechanics. The history of the theory of BEC is very interesting, and is nicely described in the biographies of Einstein[13] and London[14] and reviewed by Griffin.[15] For instance, Einstein made his predictions before quantum theory had been fully developed, and before the differences between bosons and fermions had been revealed.[16] After Einstein, important contributions were made by, most notably, London, Landau, Tisza, Bogoliubov, Penrose, Onsager, Feynman, Lee, Yang, Huang, Beliaev and Pitaevskii. An important issue has always been the relationship between BEC and superfluidity in liquid helium, an issue that was highly controversial between London and Landau (see Ref. 14). Works by Bogoliubov, Beliaev, Griffin and others showed that Bose–Einstein condensation gives the microscopic picture behind Landau's "quantum hydrodynamics". BEC is closely related to superconductivity, which can be described as being due to Bose–Einstein condensation of Cooper pairs. Thus Bose–Einstein condensation is at the heart of several macroscopic quantum phenomena.

BEC is unique in that it is a purely quantum-statistical phase transition, i.e. it occurs even in the absence of interactions. Einstein described the transition as condensation "without attractive forces".[16] This makes BEC an important paradigm of statistical mechanics, which has been discussed in a variety of contexts in condensed-matter, nuclear, particle and astrophysics.[17] On the other hand, real-life particles will always interact, and even the weakly-interacting Bose gas behaves qualitatively differently from the ideal Bose gas.[18] It was believed for quite some time that interactions would always lead to "ordinary" condensation (into a solid) before Bose–Einstein condensation would happen. Liquid helium was the only counter-example, where the light mass and concomitant large zero-point kinetic energy prevents solidification even at zero kelvin. Erwin Schrödinger wrote in 1952 in a textbook on thermodynamics about BEC: "The densities are so high and the temperatures so low — those required to exhibit a noticeable departure [from classical statistics] — that the van der Waals corrections are bound to coalesce with the possible effects of degeneration, and there is little prospect of ever being able to separate the two kinds of effect".[19] What he didn't consider were dilute systems in a metastable gaseous phase!

The quest to realize BEC in a dilute weakly interacting gas was pursued in at least three different directions: liquid helium, excitons and atomic gases. Experimental[20,21] and theoretical work[22] showed that the onset of superfluidity for liquid helium in Vycor has features of dilute-gas Bose–Einstein condensation. At sufficiently low coverage, the helium adsorbed on the porous sponge-like glass behaved like a dilute three-dimensional gas. However, the interpretation of these results is not unambiguous.[23]

Excitons, which consist of weakly-bound electron-hole pairs, are composite bosons. The physics of excitons in semiconductors is very rich and includes the formation of an electron-hole liquid and biexcitons. As nicely discussed in Refs. 24 and 25, there are systems where excitons form a weakly interacting gas. However, the initial evidence for Bose–Einstein condensation in Cu_2O (Ref. 26) was retracted.[27] Recent work in coupled quantum-well structures is very promising.[28] When excitons strongly interact with light in a cavity, they form polaritons. In such polariton systems, stimulated scattering and non-equilibrium condensates have been observed recently.[29–31]

3. Spin-Polarized Hydrogen

Dilute atomic gases are distinguished from the condensed-matter systems discussed above by the absence of strong interactions. Interactions at the density of a liquid or a solid considerably modify and complicate the nature of the phase transition. Hecht,[32] and Stwalley and Nosanow[33] used the quantum theory of corresponding states to conclude that spin polarized hydrogen would remain gaseous down to zero temperature and should be a good candidate to realize Bose–Einstein condensation in a dilute atomic gas. These suggestions triggered several experimental efforts, most notably by Silvera and Walraven in Amsterdam, by Greytak and Kleppner at MIT, and by others at Moscow, Turku, British Columbia, Cornell, Harvard, and Kyoto. The stabilization of a spin-polarized hydrogen gas[34,35] created great excitement about the prospects of exploring quantum-degenerate gases. Experiments were first done by filling cryogenic cells with the spin-polarized gas and by compressing it, and since 1985, by magnetic trapping and evaporative cooling. BEC was finally accomplished in 1998 by Kleppner, Greytak and collaborators.[36] See Refs. 9, 37–39 and in particular Ref. 40 for a full account of the pursuit of Bose–Einstein condensation in atomic hydrogen. Evidence for a phase transition in two dimensions was reported in 1998.[41]

The work in alkali atoms is based on the work in spin-polarized hydrogen in several respects:

- Studies of spin-polarized hydrogen showed that systems can remain in a metastable gaseous state close to BEC conditions. The challenge was then to find the window in density and temperature where this metastability is sufficient to realize BEC.
- Many aspects of BEC in an inhomogeneous potential,[42–44] and the theory of cold collision processes (see e.g. Ref. 45) developed in the '80s for hydrogen could be applied directly to the alkali systems.
- The technique of evaporative cooling was developed first for hydrogen[7,46] and then used for alkali atoms.

4. Laser Cooling

Laser cooling opened a new route to ultralow temperature physics. Laser cooling experiments, with room temperature vacuum chambers and easy optical access, look very different from cryogenic cells with multi-layer thermal shielding around them. Also, the number of atomic species that can be studied at ultralow temperatures was greatly extended from helium and hydrogen to all of the alkali atoms, metastable rare gases, several earth-alkali atoms, and others (the list of laser cooled atomic species is still growing). A full account of the relevant laser cooling techniques and their development is given in Refs. 47–49 and in the 1997 Nobel lectures of Chu, Cohen-Tannoudji and Phillips.[4–6]

Some papers and proposals written in the early and mid '80s, before and during the developments of the basic cooling and trapping techniques, listed quantum degeneracy in a gas as a visionary goal for this new emerging field.[50–52] However, major limitations of laser cooling and trapping were soon identified. Although there is no fundamental low temperature limit, the final temperature provided by polarization gradient cooling — about ten times the recoil energy — was regarded as a practical limit. Sub-recoil laser cooling techniques, especially in three dimensions, were harder to implement, and required long cooling times. The number and density of atoms were limited by inelastic, light-induced collisions (leading to trap loss[53,54]) and by absorption of scattered laser light,[55] which results in an outward radiation pressure (weakening the trapping potential and limiting the density). Furthermore, since the lowest temperatures could not be achieved at the highest densities,[56–58] most trapping and cooling techniques reached a maximum phase-space density $n\lambda_{\mathrm{dB}}^3 = 10^{-5}$; a value of 2.612 is needed for BEC. This was the situation when the author joined the field of cold atoms in 1990. It was only more recently that major increases in phase-space density were achieved by laser cooling,[59–61] but so far laser cooling by itself has not been able to reach BEC.

5. The Effort at MIT 1990–1996

5.1. *Improving laser cooling*

When I teamed up with Dave Pritchard at MIT in 1990 as a postdoc, the initial goal was to build an intense source of cold atoms to study cold collisions and pure long-range molecules. However, Dave and I frequently talked about the limitations in density and temperature of the current techniques and tried to develop ideas on how to get around them. One limitation of magnetic traps is that they can hold atoms only in weak-field seeking hyperfine states. Therefore, a collision between two trapped atoms can lead to a spinflip, and the Zeeman energy is converted into kinetic energy (dipolar relaxation). This process has been a major limitation to the experiments in atomic hydrogen.

First, we asked ourselves if the inclusion of electric and gravitational fields would allow the stable confinement of atoms in their lowest hyperfine states — but the answer was negative.[62] One loophole was time-dependent magnetic fields, and building on an earlier proposal,[63] I designed an experiment to confine sodium atoms with AC magnetic fields which looked feasible. However, we learnt that Eric Cornell at Boulder had developed a similar idea and experimentally implemented it[64] — so we left the idea on the drawing board. It wasn't the last time that Eric and I would develop similar ideas independently and almost simultaneously!

Trapping atoms in the lowest hyperfine state was not necessary to accomplish BEC. Already in 1986, Pritchard correctly estimated the rate constants of elastic and inelastic collisions for alkali atoms.[52] From these estimates one could easily predict that for alkali atoms, in contrast to hydrogen, the so-called good collisions (elastic collisions necessary for the evaporation process) would clearly dominate over the so-called bad collisions (inelastic two- and three-body collisions); therefore, evaporative cooling in alkalis would probably not be limited by intrinsic loss and heating processes. However, there was pessimism[65] and skepticism, and the above-mentioned experimental[64] and theoretical[62] work on traps for strong-field seeking atoms has to be seen in this context.

In those years, there were some suggestions that time-dependent potentials could lead to substantial cooling, but we showed that this was not possible.[66] Real cooling needs an open system which allows entropy to be removed from the system — in laser cooling in the form of scattered photons, in evaporative cooling in the form of discarded atoms. Dave and I brainstormed about novel laser cooling schemes. In 1991, at the Varenna summer school, Dave presented a new three-level cooling scheme.[67] Inspired by these ideas, I developed a scheme using Raman transitions. Replacing the six laser beams in optical molasses by counterpropagating beams driving the Doppler-sensitive Raman transition, we hoped to realize Doppler molasses with a linewidth that was proportional to the optical pumping rate, and therefore adjustable. We had started setting up radio frequency (RF) electronics and magnetic shields for Raman cooling when we heard that Mark Kasevich and Steve Chu were working on Raman cooling using laser pulses.[68] For this reason, and also because around the same time we had developed the idea for the Dark SPOT (spontaneous force optical trap; see later in this Section) trap, we stopped our work on Raman cooling.

Our experimental work in those years focused first on generating a large flux of slow atoms. In my first months at MIT when I overlapped with Kris Helmerson and Min Xiao, we built a sodium vapor cell magneto-optical trap (MOT). The idea was inspired by the Boulder experiment,[69] and our hope was to vastly increase the loading rate by additional frequencies or frequency chirps added to the red side of the D_2 resonance line. The idea failed — we first suspected that nearby hyperfine levels of sodium may have adversely interfered, but it was later shown that it didn't work for cesium either[70] because of the unfavorable duty cycle of the chirp. Still, except for a cryogenic setup which was soon abandoned, it was the first magneto-optical trap built at MIT (Dave Pritchard's earlier work on magneto-optical trapping was carried out at Bell Labs in collaboration with Steve Chu's group). We (Michael Joffe, Alex Martin, Dave Pritchard and myself) then put our efforts on beam slowing, and got distracted from pursuing Zeeman slowing by the idea of isotropic light slowing.[71] In this scheme, atoms are sent through a cavity with diffusely reflecting walls and exposed to an isotropic light field. For red-detuned light the atoms preferentially absorb light from a forward direction and are slowed. The experiment worked very well and it was a lot of fun to do. However, the requirements for laser power and the velocity capture range of this method were inferior to Zeeman slowing, so we decided to build an optimized Zeeman slower.

We adopted the new design by Greg Lafyatis where the magnetic field increases rather than decreases as in a conventional Zeeman slower.[72] We realized that at the magnetic field maximum it would be possible to apply some additional transverse laser cooling to collimate the slow beam. Michael Joffe, a graduate student, wound a solenoid which had radial access for four extra laser beams. The collimation worked,[73] but not as well as we had hoped, and we felt that the small gain was not worth the added complexity. Still, even without collimation, our Zeeman slower provided one of the largest slow-atom fluxes reported until then, and soon after we had a magneto-optical trap with a large cloud of sodium atoms. In hindsight, I am amazed at how many different schemes we considered and tried out, but this may have been necessary to distill the best approach.

The 1991 Varenna summer school on laser cooling was memorable to me for several reasons. I had joined the field of cold atoms just a year earlier, and there I met many colleagues for the first time and established long-lasting relationships. I still have vivid memories of one long afternoon where Dave Pritchard and I sat outside the meeting place, which offered a spectacular view on Lake Como, and brainstormed about the big goals of our field and how to approach them. Dave's encouragement was crucial to me and helped to increase my self-confidence in my new field of research. We considered options and strategies on how to combine laser cooling and evaporative cooling, something which had been on our mind for some time.

Following the example of the spin-polarized hydrogen experiment at MIT,[7] evaporation could be done in a magnetic trap using RF induced spin-flips, as suggested by Pritchard and collaborators in 1989.[74] Magnetic traps and laser cooling had already been used simultaneously in the first experiments on magnetic trapping at NIST[75] and MIT,[76] and on Doppler cooling of magnetically trapped atoms at MIT.[74,77] In 1990, a magnetic trap was loaded from a magneto-optical trap and optical molasses in Boulder.[69] The laser cooling route to BEC was summarized by Monroe, Cornell and Wieman in Ref. 78. So most of the pieces to get to BEC were known in 1990, but there was doubt about whether they would fit together.

Laser cooling works best at low densities where light absorption and light induced collisions are avoided, whereas evaporative cooling requires a high collision rate and high density. The problem is the much higher cross section for light scattering of $\sim 10^{-9}$ cm^2, while the cross section for elastic scattering of atoms is a thousand times smaller. In hindsight, it would have been sufficient to provide tight magnetic compression after laser cooling and an extremely good vacuum to obtain a lifetime of the sample that is much longer than the time between collisions, as demonstrated at Rice University.[79] However, our assessment was that one major improvement had to be done to laser cooling to bridge the gap in density between the two cooling schemes. Dave and I discussed possibilities on how to circumvent the density-limiting processes in magneto-optical traps. We considered coherent-population trapping schemes where atoms are put into a coherent superposition state which does not absorb the light. We developed some ideas on how atoms near the center of the trap would be pumped into such a dark state, but the numbers were not too promising. A few months later, a simple idea emerged. If the so-called repumping beam of the magneto-optical trap would have a shadow in the center, atoms would stay there in the lower hyperfine state and not absorb the trapping light, which is near-resonant for atoms in the upper hyperfine state. In a MOT, the density is limited by losses due to excited-state collisions and by multiple scattering of light, which results in an effective repulsive force between atoms. When atoms are kept in the dark, the trapping force decreases by a factor which is proportional to the probability of the atoms to be in the resonant hyperfine state. However, the repulsive force requires both atoms to be resonant with the light and decreases with the square of this factor. Therefore, there is a net gain in confinement by keeping atoms in the dark. Of course, there is a limit to how far you can push this concept, which is reached when the size of the cloud is no longer determined by the balance of trapping and repulsive forces, but by the finite temperature of the cloud.

The gain in density of this scheme, called Dark SPOT, over the standard MOT is bigger when the number of trapped atoms is large. So in 1992, we tweaked up the MOT to a huge size before we implemented the idea. It worked almost immediately, and we got very excited about the dark shadows cast by the trapped atoms when they were illuminated by a probe beam. We inferred[80] that the probe light had been attenuated by a factor of more than e^{-100}. This implied that we had created a cloud of cold atoms with an unprecedented combination of number and density.

5.2. *Combining laser cooling and evaporative cooling*

The following weeks and months were quite dramatic. What should we do next? Dave Pritchard had planned to use this trap as an excellent starting point for the study of cold collisions and photoassociation — and indeed other groups had major successes along these lines.[81,82] But there was also the exciting prospect of combining laser cooling with evaporative cooling. We estimated[80] the elastic collision rate in the Dark SPOT trap to be around 100 Hz which appeared to be more than sufficient to start runaway evaporation in a magnetic trap. After some discussions, the whole group decided to go for the more ambitious and speculative goal of evaporative cooling. It was one of those rare moments where suddenly the whole group's effort gets refocused. Even before we wrote the paper on the Dark SPOT trap, we placed orders for essential components to upgrade our experiment to ultrahigh vacuum and to magnetic trapping. All resources of the lab were now directed towards the evaporative cooling of sodium. The Dark SPOT trap was a huge improvement towards combining high atom number and high density in laser cooling. It turned out to be crucial to the BEC work both at Boulder[83] and at MIT[84] and seems to be still necessary in all current BEC experiments with sodium, but not for rubidium.

The next step was the design of a tightly confining magnetic trap. We decided to use the spherical quadrupole trap, which simply consists of two opposing coils — this design was used in the first demonstration of magnetic trapping.[75] We knew that this trap would ultimately be limited by Majorana flops in the center of the trap where the magnetic field is zero. Near zero magnetic field, the atomic spin doesn't precess fast enough to follow the changing direction of the magnetic field — the result is a transition to another Zeeman sublevel which is untrapped leading to trap loss. We estimated the Majorana flop rate, but there was some uncertainty about the numerical prefactor. Still, it seemed that Majorana flops would only become critical after the cloud had shrunk due to evaporative cooling, so they shouldn't get in the way of demonstrating the combination of laser cooling and evaporative cooling. After Michael Joffe presented our approach with the quadrupole trap at the QELS meeting in 1993, Eric Cornell informed me that he had independently arrived at the same conclusion. In 1993, my group reported at the OSA meeting in Toronto the transfer of atoms from the Dark SPOT trap into a magnetic trap, and the effects of truncation of the cloud using RF induced spinflips.[85]

At about this time, I joined the MIT faculty as assistant professor. Dave Pritchard made the unprecedented offer that if I stayed at MIT he would hand over to me the running lab, including two grants. To make sure that I would receive the full credit for the work towards BEC, he decided not to stay involved in a field he had pioneered and gave me full responsibility and independence. Dave told me that he wanted to focus on his other two experiments, the single-ion mass measurement and the atom interferometry, although what he gave up was his "hottest" research activity. Even now, I am moved by his generosity and unusual mentorship. The two graduate students on the project, Ken Davis and Marc-Oliver Mewes, who had started their PhD's in 1991 and 1992, respectively, deliberated whether they should stay with Dave Pritchard and work on one of his other experiments, or to continue their work on BEC in a newly-formed group headed by a largely unknown assistant professor. They both opted for the latter and we could pursue our efforts without delay, along with Michael Andrews, who joined the group in the summer of 1993.

For a few months we got distracted from our goal of evaporative cooling. Our optical molasses temperatures were higher than those reported by the NIST group,[86] and we felt that we had to learn the state of the art before we could advance to even lower temperatures. We suspected that the higher density of atoms played a role, but we had to improve our technique of temperature measurements. Our goal was to characterize the interplay of parameters in "dark" molasses where most of the atoms are pumped into the dark hyperfine state. It was also a good project for the graduate students to hone their skills and develop independence. After a few months we had made some progress, but I became concerned about the delay and the competition from Boulder. We decided to drop the project and resume our work on evaporative cooling. Up to the present day, we have never implemented accurate diagnostics for the temperature obtained in laser cooling — it was just not important.

In the spring of 1994, we saw first evidence for an increase in phase-space density by evaporative cooling. We reported these results at an invited talk at the International Quantum Electronic Conference (IQEC) in May 1994. At the same meeting, the Boulder group reported similar results and the limitations due to the Majorana flops as the temperature was reduced. It was clear that the next step was an improvement of the magnetic trap, to trap atoms at a finite bias field which would suppress the Majorana flops. During the meeting, I came up with the idea of plugging the hole with a focused laser beam: a blue-detuned laser beam focused onto the zero-magnetic field point would exert repulsive dipole forces onto the atoms and keep them away from this region (Fig. 3). This idea seemed so obvious to me that I expected the Boulder group to come up with something similar. It was only at the next conference (ICAP 1994) in Boulder,[87] when I presented our approach, that I learnt about Eric Cornell's idea of suppressing Majorana flops with a rapidly rotating magnetic field — the so-called TOP trap.[88] However, we didn't implement the optical plug immediately. We wanted first to document our observation of evaporative cooling. We realized that our fluorescence diagnostics was inadequate and implemented absorption imaging which is now the standard technique for observing Bose–Einstein condensation. In those days, we focused on direct imaging of the trapped cloud (without ballistic expansion), and Michael Andrews and Marc-Oliver Mewes developed a sophisticated computer code to simulate absorption images in inhomogeneous magnetic fields. We thought that this would be a useful tool, but we rapidly advanced to much lower temperatures where the inhomogeneous Zeeman shifts were smaller than the linewidth, and never needed the code again after our first paper on evaporative cooling.[89]

In late 1994, we had a "core melt down". The magnetic trap was switched on without cooling water, and the silver solder joints of the coils melted. Since in those days the magnetic coils were mounted inside the vacuum chamber, we had a catastrophic loss of vacuum and major parts of our setup had to be disassembled. I will never forget the sight of coils dripping with water behind a UHV viewport. This happened just a few hours before MIT's president, Charles Vest, visited our lab to get first-hand information on some of the research done on campus. He still remembers this event. We had lost weeks or months of work in a very competitive situation. I was despondent and suggested to the group that we go out for a beer and then figure out what to do, but the students immediately pulled out the wrenches and started the repair. I was moved to see their dedication and strength, even at this difficult time. We replaced the magnetic trap by a much sturdier one. This turned out to be crucial for the implementation of the plugged trap where the precise alignment of a laser beam relative to the magnetic field center was important. So on hindsight the disaster may not have caused a major delay.

In early 1995, I had to tell my three graduate students that we were rapidly using up start-up money and urgently needed one of our two pending proposals approved. Otherwise we would not be able to continue spending money in the way we had done until then and would slow down. Fortunately, in April 1995, the NSF informed me that my proposal was funded. It is interesting to look at some of the reviewers comments now, seven years later: "It seems that vast improvements are required [in order to reach BEC]... the current techniques are so far from striking range for BEC that it is not yet possible to make... an assessment..."; "The scientific payoffs, other than the importance of producing a BEC itself, are unclear". And a third reviewer: "...there have been few specific (or realistic) proposals of interesting experiments that could be done with a condensate". Despite the skepticism, all reviewers concluded that the proposed "experiments are valuable and worth pursuing". After we received the funding decision, the whole group celebrated with dinner, and a fourth graduate student (Dallin Durfee), who had expressed his interest already months earlier, could finally be supported.

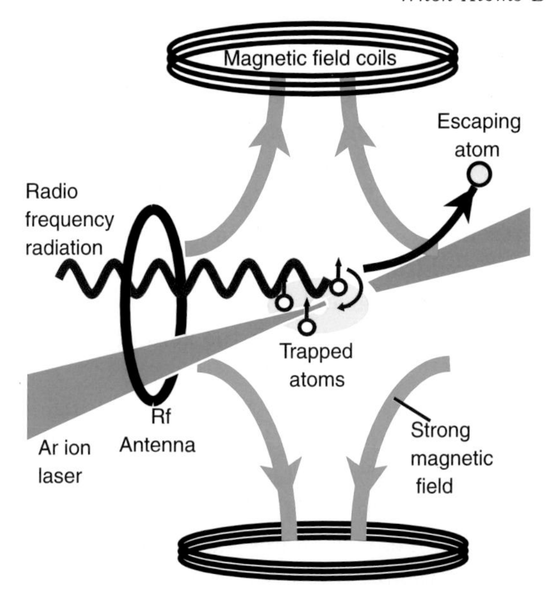

Fig. 3. Experimental setup for cooling atoms to Bose–Einstein condensation. Sodium atoms are trapped by a strong magnetic field, generated by two coils. In the center, the magnetic field vanishes, which allows the atoms to spin-flip and escape. Therefore, the atoms are kept away from the center of the trap by a strong (3.5 W) argon ion laser beam ("optical plug"), which exerts a repulsive force on the atoms. Evaporative cooling is controlled by radio-frequency radiation from an antenna. The RF selectively flips the spins of the most energetic atoms. The remaining atoms rethermalize (at a lower temperature) by collisions among themselves. Evaporative cooling is forced by lowering the RF frequency.

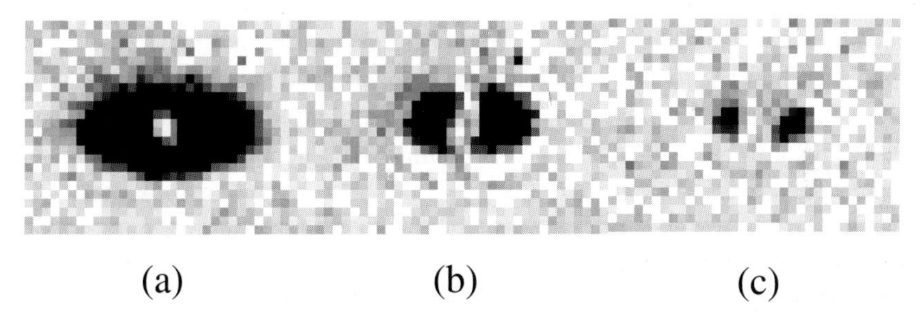

Fig. 4. Absorption images of atom clouds trapped in the optically plugged trap. Cloud (a) is already colder than was attainable without the "plug" (Ar ion laser beam). Cloud (b) shows the break-up of the cloud into two "pockets" in the two minima of the potential. The size of cloud (c) reached the optical resolution of the imaging system ($\leq 10\ \mu$m) still absorbing 90% of the probe light. This sets an upper bound on temperature ($\leq 10\ \mu$K) and a lower bound on density (5×10^{12} cm^{-3}).

In late December 1994, our paper on evaporative cooling was submitted, and we were free to focus on plugging the hole. We had to learn how to align a powerful argon ion laser beam and image it through many attenuators without major distortions. When the plug was aligned, the result was spectacular (Fig. 4). We could immediately cool down to lower temperatures and keep many more atoms. During evaporation, the cloud became so cold and small that we could not resolve it any more. The highest phase space density measured was a factor of thirty below BEC, but we may have been even closer. We had only a few runs of the experiment before we ran into severe vacuum problems. We focused initially on spatial imaging and became limited by resolution, whereas ballistic expansion and time-of-flight imaging would not have suffered from this limitation. We also thought that BEC would be accomplished at lower densities and in larger clouds, so we worked on adiabatic decompression and ran into problems with the zero of the magnetic field moving away from the plug.

In those months, we were plagued by vacuum problems. The coils inside the vacuum showed some strange outgassing behavior and the vacuum slowly deteriorated. We went through several bakeouts of the ultrahigh-vacuum chamber in the spring and summer of 1995. Furthermore, Ken Davis had to write his PhD thesis and stopped working in the lab. It is interesting to recall my assessment of the field in those months; I didn't realize that BEC was just around the corner. In Tom Greytak's and Dan Kleppner's group the BEC transition was approached to within a factor of 3.5 in temperature in 1991,[90] but it took several more years to advance further. So I prepared for a long haul to cover the last order of magnitude to BEC.

By this time, the group was reinforced by Dan Kurn (now Dan Stamper-Kurn), a graduate student, and Klaasjan van Druten, my first postdoc. After months of working on vacuum and other problems, we were just ready to run the machine again when we heard about the breakthrough in Boulder in June of 1995. We feverishly made several attempts with traps plugged by focused laser beams and light sheets, and tried different strategies of evaporation without success. The clouds disappeared when they were very cold. We conjectured that some jitter of the laser beam was responsible, and when accelerometers indicated vibrations of our vacuum chambers, we immediately decided to eliminate all turbo and mechanical pumps. Unfortunately, when we were exchanging the turbo pump on our oven chamber against an ion pump, we caused a leak in the ultrahigh vacuum part and had to go through another long bake-out. We also implemented a pointing stabilization for the optical plug beam. But when we finally obtained BEC, we realized that it didn't improve the cooling.

These were difficult months for me. The Rice group had cooled lithium to quantum degeneracy.[79] A new subfield of atomic physics was opening up, and I was afraid that our approach with sodium and the plugged trap would not be successful and we would miss the excitement. I considered various strategies. Several people suggested that I adopt the successful TOP trap used at Boulder. But I had already started to study several possible configurations for magnetic confinement. I realized that a highly elongated Ioffe–Pritchard trap with adjustable bias field could provide a good confinement that was equivalent or superior to the TOP trap. Around August 1995, Dan Kurn worked out an optimized configuration, which was the cloverleaf winding pattern.[91] I considered having the whole group work on this new approach, but several in my group wanted to give the plugged trap a few more attempts and at least characterize how far we could approach BEC with our original approach. Fortunately, we followed that suggestion — it is always a good idea to listen to your collaborators.

5.3. *BEC in sodium*

This was the situation on September 29, 1995, when we observed BEC in sodium for the first time. The goal of the run was to measure the lifetime of the trapped atoms and characterize possible heating processes. For our ultrahigh vacuum pressure, rather slow evaporation should have been most efficient, but we found out that faster evaporation worked much better. This was a clear sign for some other loss

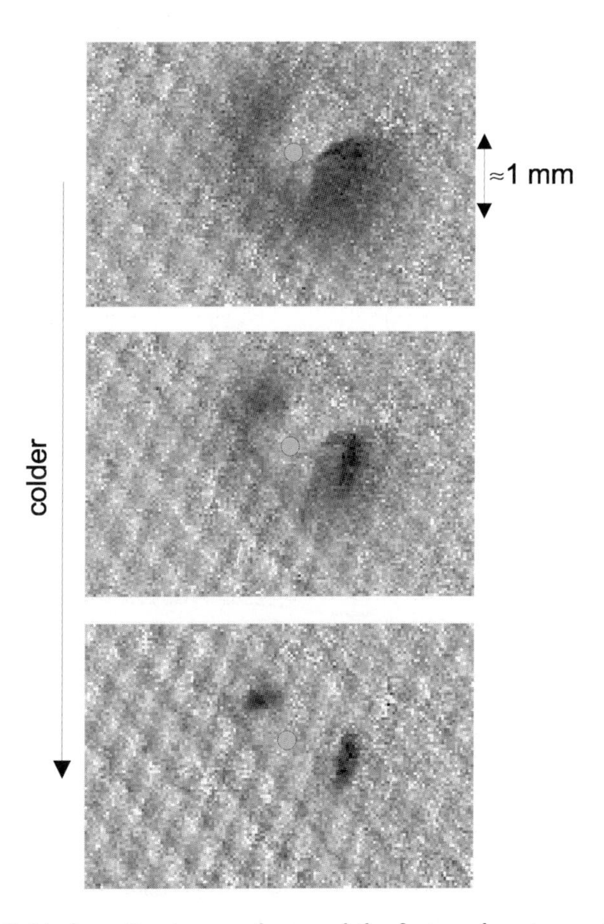

Fig. 5. Time-of-flight absorption images of some of the first condensates produced at MIT in the night of September 29, 1995. After the magnetic quadrupole trap was switched off, the atom cloud expanded ballistically. However, since the optical plug (indicated by black circles) could not be turned off at the same time, it distorted the expanding cloud. Still, as the temperature was lowered from top to bottom, a distinctly sharp shadow appeared marking the presence of a condensate.

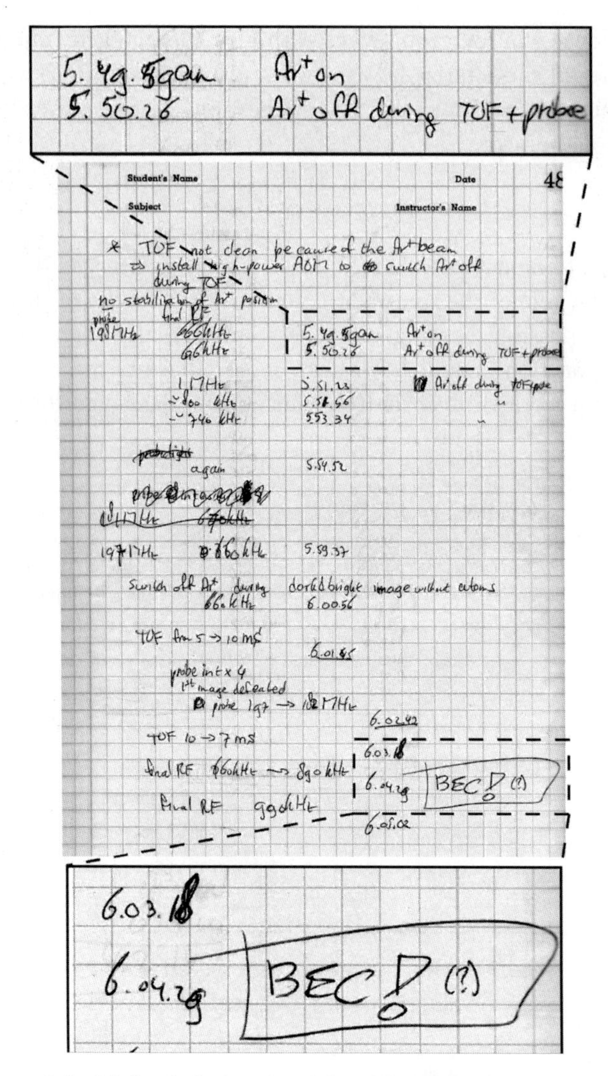

Fig. 6. One page of the lab book during the night of September 29, 1995, when BEC was first observed at MIT. The handwriting is by Klaasjan van Druten. At 5:50 a.m., we had installed a new acousto-optical modulator to switch-off the optical plug (Ar ion laser beam). Fifteen minutes later, we had the first definitive evidence for BEC in sodium.

or heating process, e.g. due to fluctuations in the position of the plug. Around 11:30 p.m., an entry in the lab book states that the lifetime measurements were not reliable, but they indicated lifetimes around ten seconds, enough to continue evaporation. A few minutes later we saw some dark spots in time-of-flight absorption images, but they were quite distorted since the optical plug beam, which we couldn't switch off, pushed atoms apart during the ballistic expansion (Fig. 5). Still, the sudden appearance of dark spots meant groups of atoms with very small relative velocity. For the next few hours, we characterized the appearance of those spots,

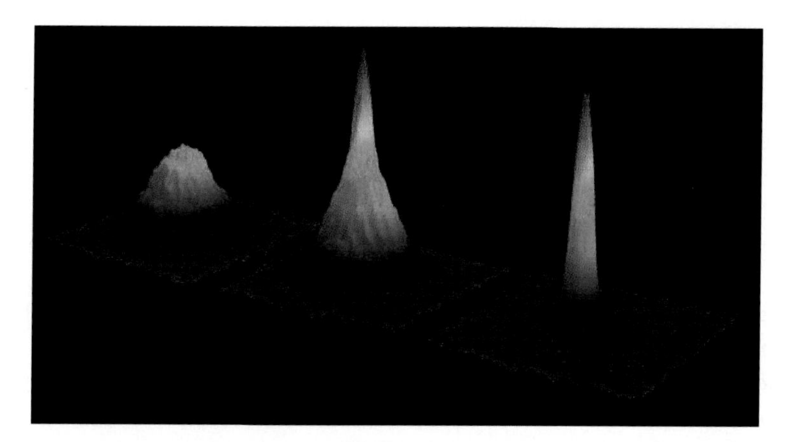

Fig. 7. Observation of Bose–Einstein condensation by absorption imaging. Shown is absorption versus two spatial dimensions. The Bose–Einstein condensate is characterized by its slow expansion observed after 6 ms time-of-flight. The left picture shows an expanding cloud cooled to just above the transition point; middle: just after the condensate appeared; right: after further evaporative cooling has left an almost pure condensate. The total number of atoms at the phase transition is about 7×10^5, the temperature at the transition point is 2 μK.

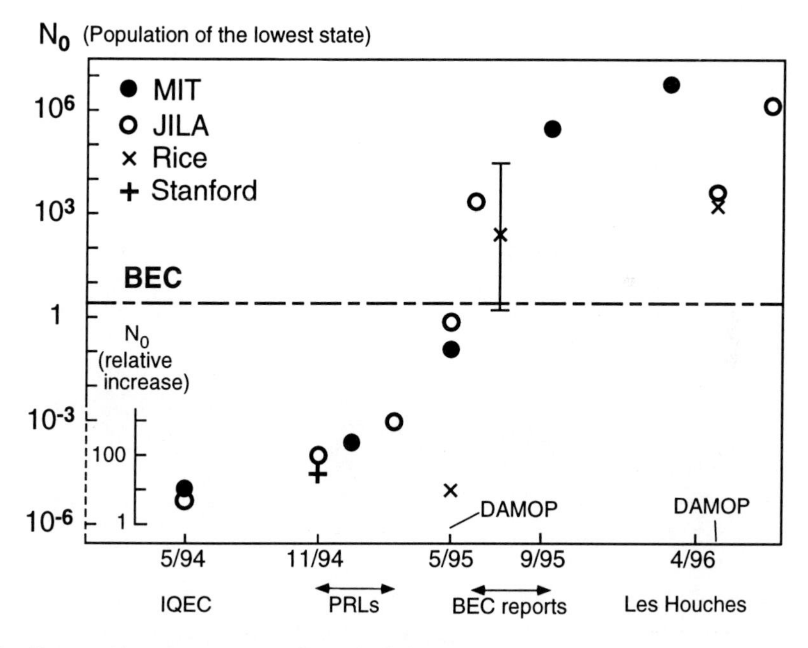

Fig. 8. Progress in evaporative cooling of alkali atoms up to 1996. The number of atoms in the lowest quantum state is proportional to the phase-space density, and has to exceed a critical number of 2.612 to achieve Bose–Einstein condensation. For $N_0 < 10^{-3}$, the increase in phase-space density due to evaporation is plotted. For the Rice result of July 1995 see Ref. 79 and the erratum.[93]

but then decided that further progress required an acousto-optical modulator to switch off the optical plug. Between 4:00 and 5:30 in the early morning, we installed optics and RF electronics and were finally able to switch off the argon ion laser beam during ballistic expansion. Fifteen minutes later, we observed the bimodal distributions that are now the hallmark of BEC. The lab book of this night captured the excitement of the moment (Fig. 6).

Those first measurements were done by imaging the atoms in the lower hyperfine ($F = 1$) state. For the next run, which took place a few days later, we prepared optical pumping and imaging on the cycling $F = 2$ transition, and obtained a much better signal-to-noise ratio in our images. The occurrence of BEC was very dramatic (Fig. 7). Our animated rendering of the data obtained in that run (done by Dallin Durfee) became well-known (see Ref. 92). We had obtained condensates with 500,000 atoms, 200 times more than in Boulder, with a cooling cycle (of only nine seconds) 40 times shorter. Our paper was quickly written and submitted only two weeks after the experiment.

In my wildest dreams I had not assumed that the step from evaporative cooling to BEC would be so fast. Figure 8 shows how dramatic the progress was after laser and evaporative cooling were combined. Within less than two years, the number of alkali atoms in a single quantum state was increased by about twelve orders of magnitude — a true singularity demonstrating that a phase transition was achieved!

MIT with its long tradition in atomic physics was a special place to pursue the BEC work. The essential step was the combination of laser cooling and evaporative cooling. My next-door neighbors in Building 26 at MIT have been Dave

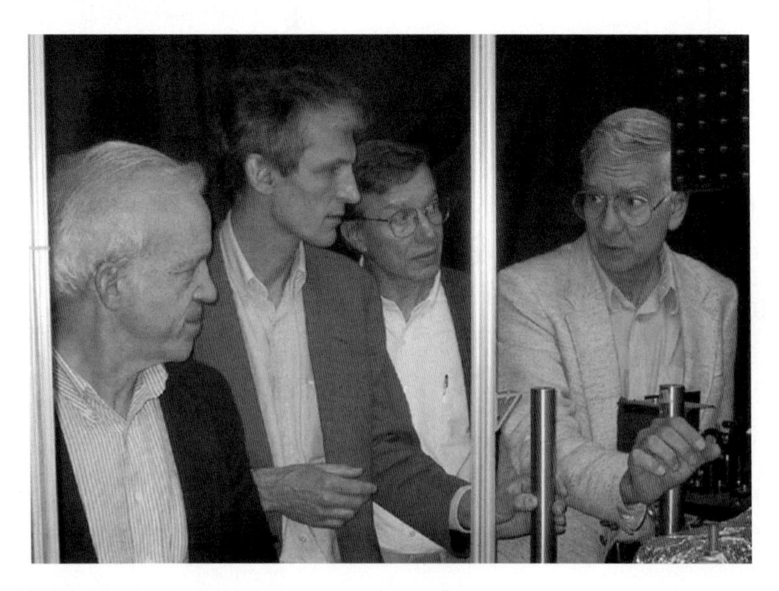

Fig. 9. MIT faculty in ultralow-temperature atomic physics. Dan Kleppner, W. K., Tom Greytak and Dave Pritchard look at the latest sodium BEC apparatus.

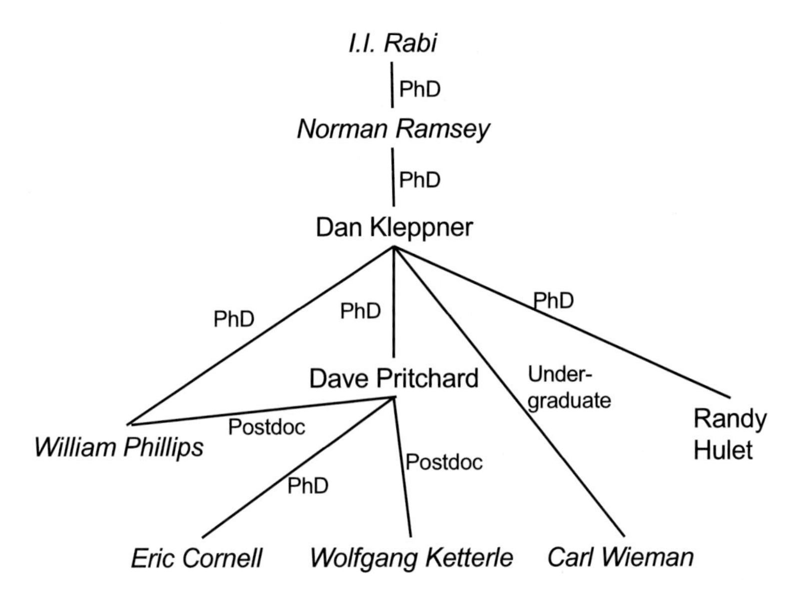

Fig. 10. Family tree of atomic physicists. People with name in italics are Nobel laureates.

Pritchard, a pioneer in laser cooling who conceived the magneto-optical trap, and Dan Kleppner, who together with Harald Hess and Tom Greytak conceived and realized evaporative cooling (see Fig. 9). I feel privileged for the opportunity to combine their work and take it to the next level. It is hard to overestimate the roles which Dave Pritchard and Dan Kleppner have played for modern atomic physics. The family tree of atomic physicists (Fig. 10) shows some of the remarkable physicists that were trained and inspired by them.

Looking back, it seems that many techniques such as the Dark SPOT, compressed MOT,[94] the TOP trap and the optically plugged trap were critical for first demonstrating BEC, but by no means indispensable. This is best illustrated by the experiment at Rice which used only Doppler cooling to load the magnetic trap — a technique which had been developed in the '80s. The collision rate was slow, but an excellent vacuum made a very slow evaporation process possible.[79] So in hindsight, BEC in alkali gases did not require major innovations in cooling and trapping. It merely required enough optimism to risk a few years in the attempt to combine laser and evaporative cooling. Such an attempt needed a few years of very focused work as it involved the integration of several technologies that were not standard in the field, including ultrahigh vacuum, sensitive CCD cameras and image processing, high-current power supplies for magnetic traps, and flexible computer control of a multi-step cooling and detection process. Figure 11 compares a state-of-the-art laser cooling experiment in 1993 to a BEC experiment in 2001 using the same vacuum apparatus in the same laboratory at MIT. A lot of components have been added, and I continue to be impressed by my collaborators, who now handle experiments far more complex than I did some five years ago.

Fig. 11. Comparison of a laser cooling and BEC experiment. The upper photograph shows the author in 1993 working on the Dark SPOT trap. In the following years, this laser cooling experiment was upgraded to a BEC experiment. The lower photograph shows the same apparatus in 2001 after many additional components have been added.

5.4. *The cloverleaf trap*

After our first observation of BEC, we made the right decision for the wrong reason. We expected many other groups to quickly upgrade their laser cooling experiments to magnetic trapping and evaporative cooling, and to join in during the next few months. Nobody expected that it would take almost two years before the next groups succeeded in reaching BEC (the groups of Dan Heinzen, Lene Hau, Mark Kasevich and Gerhard Rempe followed in 1997). I was concerned that our plugged trap would put us at a disadvantage since the trapping potential strongly depended on the shape and alignment of the laser focus. So we decided to install the cloverleaf trap instead and discontinue our plugged trap after only two experimental BEC "runs".

Since we did not want to break the vacuum, we installed the new trap in an unfavorable geometry. The magnet coils for the plugged trap were oriented vertically in re-entrant flanges and when we replaced them with cloverleaf coils, the weakly confining axis of the Ioffe–Pritchard trap was vertical. In such a geometry, the gravitational sag would reduce the efficiency of RF induced evaporation since

Fig. 12. Experimental setup for cooling sodium atoms to Bose–Einstein condensation around 1996. The atoms are trapped and cooled in the center of the UHV chamber. The atomic beam oven and the Zeeman slower are to the left (outside the photo). The cloverleaf magnetic trap was mounted horizontally in re-entrant flanges. Only the leads for the current and water cooling are visible. The diagonal flange above accommodated a BNC feedthrough for radio frequency fields which were used to control the evaporative cooling. The lens and the mirror above the chamber were used to observe the condensate by dispersive or absorption imaging.

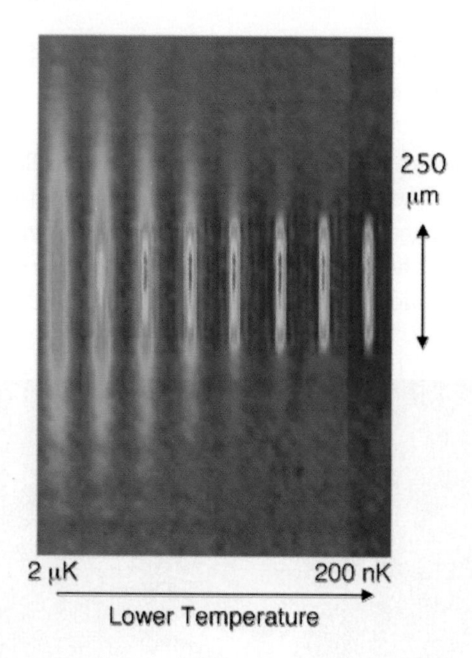

Fig. 13. Phase contrast images of trapped Bose gases across the BEC phase transition. At high temperature, above the BEC transition temperature, the density profile of the gas is smooth. As the temperature drops below the BEC phase transition, a high-density core of atoms appears in the center of the distribution. This is the Bose–Einstein condensate. Lowering the temperature further, the condensate number grows and the thermal wings of the distribution become shorter. Finally, the temperature drops to the point where a pure condensate with no discernible thermal fraction remains. Each image shows an equilibrated gas obtained in one complete trapping and cooling cycle. The axial and radial frequencies are about 17 and 230 Hz, respectively.

atoms would only evaporate at the bottom of the cloud.[8,95] But before breaking the vacuum and reorienting the coils, we wanted to see the limitation. Around December 1995, when we were just starting to look at the efficiency of evaporation, we lost the vacuum once again due to a cracked ceramic part in an electric feedthrough and decided to reorient the whole experiment, with the weakly confining axis of the trap now aligned horizontally. Since that time, now more than six years, the machine has been under vacuum. This is in sharp contrast to the conditions in 1995, when we had to open the chamber, pump down and bake out every couple of months. Finally, we had learned from our previous mistakes and developed a very systematic procedure for pump downs and bakeouts.

I still remember the night of March 13, 1996, when the experiment was up and running, and Klaasjan van Druten and I had fine-tuned the bias field of the magnetic trap, so that the switch-over to the new magnetic trap was finally completed. It was already after midnight, too late to start some serious work, when Klaasjan asked half jokingly why don't we just try to get BEC. Without knowing what our temperatures and densities were, without having ever measured the trap frequencies, we played around with the RF sweep that determines the cooling trajectory, and a condensate showed up around 2:10 a.m. We were relieved since we had not produced condensates for almost half a year, but also the ease at which we got the condensate in a new trap told us our setup was robust and that we were ready to switch from engineering cooling schemes and traps to the study of the condensate. The cloverleaf trap and other winding patterns for the Ioffe–Pritchard configuration are now used by almost all BEC experiments. Figure 12 shows the experimental setup during those days.

Why had we not considered this trap earlier and avoided the detours with the quadrupole trap, Majorana flops and plugging the hole? First, the quadrupole trap was simpler to build, and it allowed us to pursue evaporative cooling faster. Second, we initially favored the quadrupole trap based on an analysis which shows that confinement by a linear potential is much stronger than by the quadratic potential of the Ioffe–Pritchard configuration.[10] However, a very elongated Ioffe–Pritchard trap provides effectively linear confinement in the two radial directions, and it was only in 1995 that I realized that it would be easy to adiabatically deform the round laser-cooled cloud to such an elongated shape.

The next weeks were exciting and dramatic: we implemented dispersive imaging and saw for the first time the condensate in the trap. We could take images non-destructively and recorded two sequential images of the same condensate. After year-long concerns of how fragile and sensitive the condensate would be once created, it was an overwhelming experience to observe the condensate without destroying it. Figure 13 shows a spatial image of a condensate; it was taken in non-destructive dispersive imaging. We first implemented dispersive imaging using the dark-ground technique,[96] but soon upgraded to phase-contrast imaging, which was the technique used to record the figure.

In the first week of April 1996, there was a workshop on "Collective effects in ultracold atomic gases" in Les Houches, France, where most of the leading groups were represented. It was the first such meeting after the summer of 1995, and it was not without strong emotions that I reported our results. Since no other experimental group had made major progress in BEC over the last few months, it was our work which provided optimism for further rapid developments.

5.5. *Interference between two condensates*

After we got BEC in the cloverleaf trap, both the machine and the group were in overdrive. After years of building and improving, frequent failures and frustration, it was like a phase transition to a situation where almost everything worked. Within three months after getting a condensate in the cloverleaf trap we had written three papers on the new trap and the phase transition,[91] on non-destructive imaging,[96] and on collective excitations.[97] Klaasjan van Druten left the group, shortly after Christopher Townsend had joined us as a postdoc. As the next major goal, we decided to study the coherence of the condensate. With our optical plug, we had already developed the tool to split a condensate into two halves and hoped to observe their interference, which would be a clear signature of the long-range spatial coherence.

Around the same time, the idea came up to extract atoms from the condensate using RF induced spinflips — the RF output coupler. Some theorists regarded an output coupler as an open question in the context of the atom laser. I suggested to my group that we could simply pulse on the radio frequency source that was already used during evaporation, and couple atoms out of the condensate by flipping their spin to a non-trapped state (Fig. 14). The experiment worked the first time we tried it (but the quantitative work took awhile[98]). I have never regarded the output coupler as one of our major accomplishments because it was so simple, but it had impact on the community and nobody has ever since regarded outcoupling as a problem!

In July 1996, we had the first results on the RF output coupler, and also seen first fringes when two condensates were separated with a sheet of green light and overlapped in ballistic expansion. I was in Australia for vacation and for the IQEC conference in Sydney. By e-mail and telephone I discussed with my group the new results. The fringes were most pronounced when the condensates were accelerated into each other by removing the light sheet shortly before switching off the magnetic trap. We concluded that some of the fringes may be related to sound and other collective effects that occur when two condensates at fairly high density "touch" each other. I presented those results at the Sydney meeting only to illustrate we were able to do experiments with two condensates, but now we had to sort out what was happening.

It took us four more months until we observed clean interference between two condensates. When two condensates that were initially separated by a distance d interfere and the interference pattern is recorded after a time t of ballistic expansion, then the fringe spacing is the de Broglie wavelength h/mv associated with the relative velocity $v = d/t$. For our geometry with two condensates about 100 μm in length, we estimated that we would need at least 60 ms of time-of-flight to observe fringes with a 10 μm period, close to the resolution of our imaging system. Unfortunately, due to gravity, the atoms dropped out of the field of view of our windows after 40 ms. So we tried to gain a longer expansion time in a fountain geometry where we magnetically launched the atoms and observed them when they fell back through the observation region after more than 100 ms,[99] but the clouds were distorted. We also tried to compensate gravity by a vertical magnetic field gradient. Some time later I learnt about new calculations by the theory group at the Max Planck Institute in Garching, showing that the effective separation of two elongated condensates is smaller than their center-of-mass separation.[100] This meant that we could observe interference fringes after only 40 ms, just before the atoms fell out of the observation region. We immediately had a discussion in the group and decided to stop working on fountains and "anti-gravity" and simply let the atoms fall by 8 mm during 40 ms.

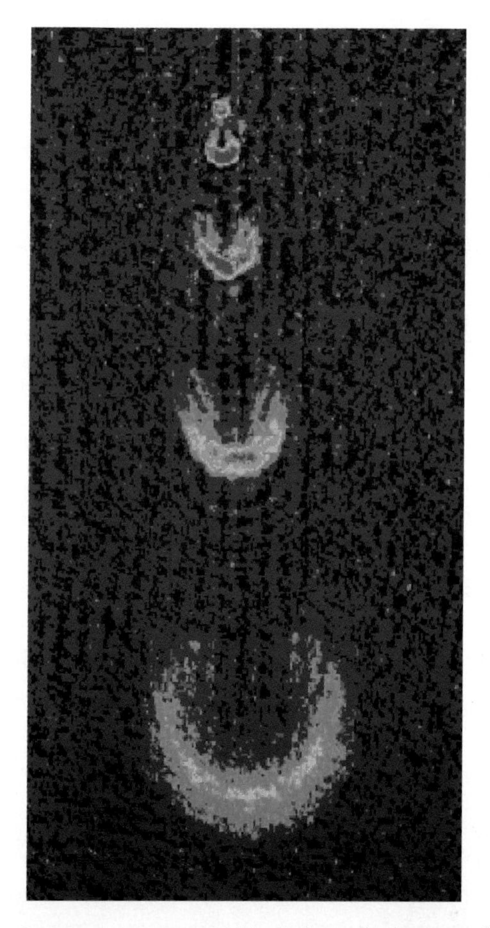

Fig. 14. The MIT atom laser operating at 200 Hz. Pulses of coherent sodium atoms are coupled out from a Bose–Einstein condensate confined in a magnetic trap (field of view 2.5×5.0 mm^2). Every 5 ms, a short RF pulse transferred a fraction of these atoms into an unconfined quantum state. These atoms were accelerated downward by gravity and spread out due to repulsive interactions. The atom pulses were observed by absorption imaging. Each pulse contained between 10^5 and 10^6 atoms.

We made some ambiguous observations where we saw low-contrast fringes together with some optical interference patterns of the probe light, but the breakthrough came on November 21, 1996, when we observed striking interference patterns (Fig. 15). I still remember the situation late that night when we wondered how could we prove beyond all doubt that these were matter-wave interference patterns and not some form of self-diffraction of a condensate confined by a light sheet and then released. We came up with the idea of eliminating one of the condensates in the last moment by focusing resonant yellow light on it. Whimsically, this laser beam was dubbed the "flame thrower". If the fringes were self-diffraction due to the sharp edge in the confinement, they would remain; if they were true interference

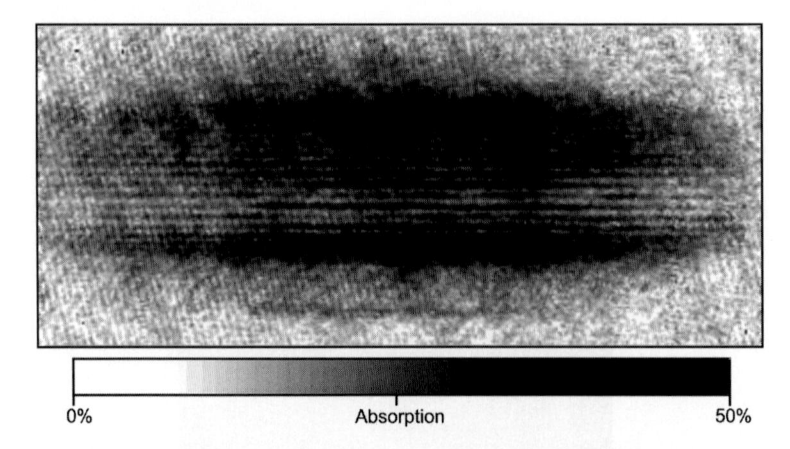

Fig. 15. Interference pattern of two expanding condensates observed after 40 ms time-of-flight. The width of the absorption image is 1.1 mm. The interference fringes have a spacing of 15 μm and are strong evidence for the long-range coherence of Bose–Einstein condensates.

they would vanish. This was like a double slit experiment in optics where you cover one of the slits. It took a few hours to align the new laser beam, and we verified in phase-contrast imaging that we were able to selectively eliminate one of the two condensates.

Fig. 16. Team photo. This photo was taken in early 1996 in front of the MIT dome. The bottle of champagne was emptied to celebrate BEC in the cloverleaf trap. Names and photos of other collaborators during the period 1992–1996 have been added.

We had a switch in our control panel which toggled between condensate elimination on and off. Then we went back and aligned the setup for the observation of interference. When we toggled the switch we had to wait for about half a minute until a new condensate was produced. This was the moment of truth. If the fringes appeared without a second condensate, then nature would have fooled us for the whole night — but they disappeared and an enormous tension disappeared, as well. It was already early the next morning, with people arriving to work. I walked to Dan Kleppner's office and told him there was something he should see. So he shared the moment with us where we toggled the switch on alternating cooling cycles and correspondingly, the interference pattern disappeared and reappeared. Interference between two light beams is quite a sight, but with atoms it is more dramatic. Destructive interference means that atoms plus atoms add up to vacuum!

The evidence for interference was so compelling that we submitted our paper based solely on the data of one experimental run.[101] This run is memorable to me for another reason: it was to be the last time I played a major role in preparing and running an experiment. During the night, I had put in the optics for the "flame thrower". Up to then, I was familiar with every piece of equipment in the lab and never thought this could change quickly, but it was like another phase transition. Hans-Joachim Miesner had just arrived, the first postdoc who stayed for more than a year, and he soon took over much responsibility for organizing the lab. There were more demands on my time to write papers and give talks, the group grew with the addition of two more graduate students (Shin Inouye and Chris Kuklewicz), and we had intensified our efforts to build a second BEC experiment. All this coincided in a few months. After earning my PhD in 1986, I spent eleven more years in the lab during three postdoc positions and as an assistant professor, but now began to play an advisory role.

The papers on the RF output coupler[98] and the interference[101] of two condensates appeared in the same week in January 1997. Together they demonstrated the ability to create multiple pulses of coherent atoms, and have been regarded as the realization of an atom laser. The period starting with the early dreams of pursuing BEC and ending with the observation of the coherence of the condensate was remarkable. It was full of speculation, dreams, unknown physics, failures and successes, passion, excitement, and frustration. This period fused together a team of very different people who had one common denominator: the passion for experimental physics. It was a unique experience for me to work with these outstanding people (Fig. 16).

6. The Magic of Matter Waves

Many studies of BECs have been performed over the last several years. The progress until 1998 is nicely summarized in the Varenna summer school proceedings.[102] The studies that were most exciting for me displayed macroscopic quantum mechanics, the wavelike properties of matter on a macroscopic scale. These were also phenomena that no ordinary gas would show, and illustrated dramatically that a new form of matter had been created. The interference of two condensates presented above (Fig. 15) is one such example. In the following, I want to discuss the amplification of atoms and the observation of lattices of quantized vortices.

These two examples are representative of the two areas into which research on gaseous BEC can be divided: in the first (which could be labelled "The atomic condensate as a coherent gas" or "Atom lasers"), one would like to have as little interaction as possible — almost like the photons in a laser. The experiments are preferably done at low densities. The Bose–Einstein condensate serves as an intense source of ultracold coherent atoms for experiments in atom optics, in precision studies or for explorations of basic aspects of quantum mechanics. The second area could be labelled "BEC as a new quantum fluid" or "BEC as a many-body system". The focus here is on the interactions between the atoms that are most pronounced at high densities. The coherent amplification of atoms is an example of atom optics with condensates, and the study of vortices addresses the superfluid properties of the gas.

6.1. *Amplification of atoms in a Bose–Einstein condensate*

Since atoms are de Broglie waves, there are many analogies between atoms and light, which consists of electromagnetic waves. This is exploited in the field of atom optics where atoms are reflected, diffracted and interfere using various atom-optical elements.[103] One important question was whether these analogies can be extended to the optical laser, which is based on the amplification of light. When our group demonstrated a rudimentary atom laser in 1997 we had solved the problem of outcoupling (or extracting) atoms from the BEC and of verifying their coherence. The atomic amplification process happened during the formation of the Bose–Einstein condensate,[104] which is quite different from the way light is amplified in passing through an active medium. It was only in 1999 that our group managed to observe the amplification of atoms passing through another cloud of atoms serving as the active medium[105] (simultaneously with the group in Tokyo[106]).

Amplifying atoms is more subtle than amplifying electromagnetic waves because atoms can only change their quantum state and cannot be created. Therefore, even if one could amplify gold atoms, one would not realize the dreams of medieval alchemy. An atom amplifier converts atoms from the active medium into an atomic wave that is exactly in the same quantum state as the input wave (Fig. 17).

The atom amplifier requires a reservoir, or an active medium, of ultracold atoms that have a very narrow spread of velocities and can be transferred to the atomic beam. A natural choice for the reservoir was a Bose–Einstein condensate. One also needs a coupling mechanism that transfers atoms from the reservoir at rest to an input mode while conserving energy and momentum. This transfer of atoms was accomplished by scattering laser light. The recoil of the scattering process accelerated some atoms to exactly match the velocity of the input atoms (Fig. 18).

Not only were the atoms amplified, but were in exactly the same motional state as the input atoms, i.e. they had the same quantum-mechanical phase. This was verified by interfering the amplified output with a copy of the input wave and observing phase coherence.

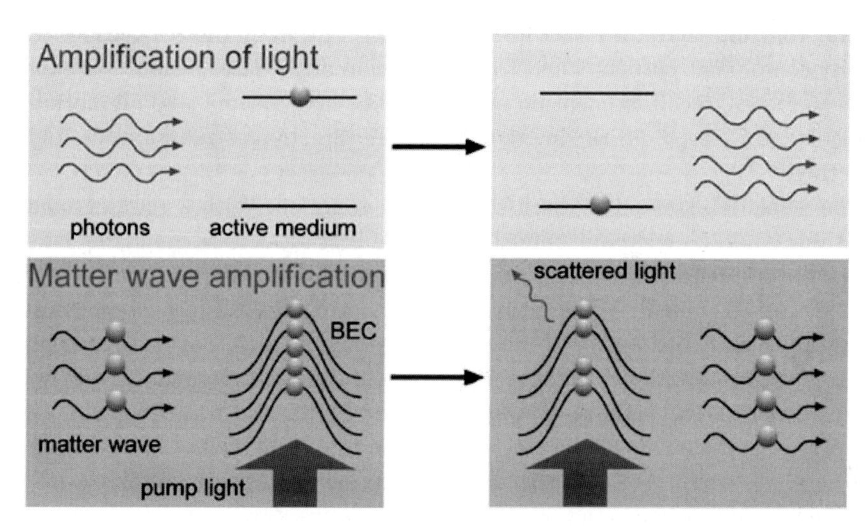

Fig. 17. Amplification of light and atoms: In the optical laser, light is amplified by passing it through an excited inverted medium. In the MIT atom amplifier, an input matter wave is sent through a Bose–Einstein condensate illuminated by laser light. Bosonic stimulation by the input atoms causes light to be scattered by the condensate exactly at the angle at which a recoiling condensate atom joins the input matter wave and augments it.

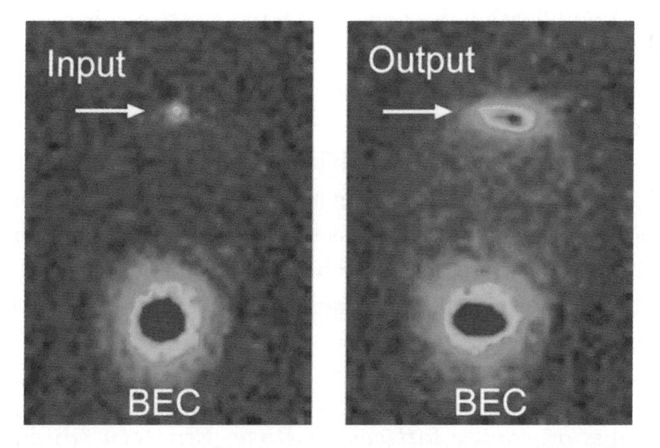

Fig. 18. Observation of atom amplification. Atom amplification is probed by sending an input beam through the atom amplifier which is a Bose–Einstein condensate (BEC) illuminated with laser light. On the left side, the input beam has passed through the condensate without amplification. Some 20 ms later, a shadow picture is taken of the condensate and the input atoms. When the amplification process was activated by illuminating the condensate with laser light, the output pulse contained many more atoms than the input pulse — typical amplification factors were between 10 and 100. The field of view is 1.9×2.6 mm^2.

This direct observation of atom amplification in the summer of 1999 was preceded by a surprising occurrence late one night in October 1998 when we discovered a new form of superradiance.[107] We were studying Bragg spectroscopy[108] and illuminated a BEC with two laser beams. I had no role in the running of the experiment and was working in my office, when around midnight the students came from the lab and told me that they saw atoms shooting out from the condensate with a velocity component perpendicular to the direction of the laser beams. We expected atoms to receive recoil momentum only along the laser beams, and all motion perpendicular to it to be diffuse due to the random direction of spontaneous Rayleigh scattering.

The whole lab started to discuss what was going on. With a running machine, everything could be tried out immediately. The first ideas were mundane: let's illuminate the condensate with only one laser beam and see what happen (the directional beams remained). We scrutinized the experimental setup for bouncing laser beams or beams which had not been completely switched off, but we found nothing. Increasingly, we considered that the observed phenomenon was genuine and not due to some experimental artifact. Knowing that the condensate was pencil-shaped, the idea of laser emission along the long condensate axis came up, and this was already very close. We decided to stop the general discussion and continue taking data; the machine was running well and we wanted to take advantage of it. So some students, including Shin Inouye and Ananth Chikkatur, characterized the phenomenon, while Dan Stamper-Kurn and I went to a black board and tried to figure out what was going on. Within the next hour, we developed the correct semi-classical description of superradiance in a condensate. In the lab, the predicted strong dependence on laser polarization was verified. A few months later we realized how we could use the superradiant amplification mechanism to build a phase-coherent atom amplifier. However, the labs were undergoing complete renovation at this point and we had to wait until the machine was running again before the phase-coherent amplification was implemented.

The demonstration of an atom amplifier added a new element to atom optics. In addition to passive elements like beam splitters, lenses and mirrors, there is now an active atom-optical element. Coherent matter wave amplifiers may improve the performance of atom interferometers by making up for losses inside the device or by amplifying the output signal. Atom interferometers are already used as precise gravity and rotation sensors.

6.2. *Observation of vortex lattices in Bose–Einstein condensates*

Quantum mechanics and the wave nature of matter have subtle manifestations when particles have angular momentum, or more generally, when quantum systems are rotating. When a quantum-mechanical particle moves in a circle the circumference of the orbit has to be an integer multiple of the de Broglie wavelength. This quantization rule leads to the Bohr model and the discrete energy levels of the

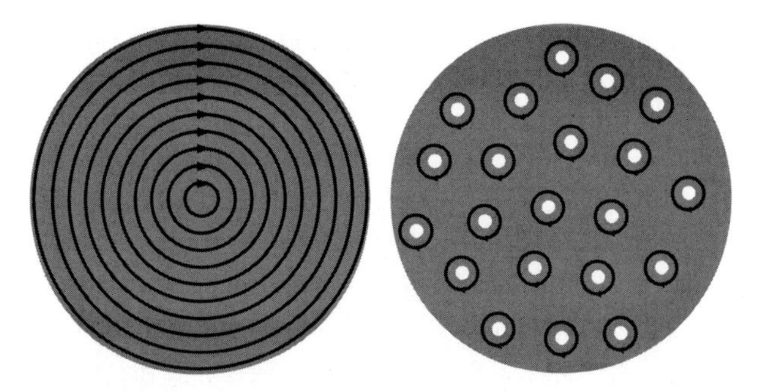

Fig. 19. Comparison of the flow fields of rotating normal liquids and superfluids. A normal fluid undergoes rigid body rotation, whereas a superfluid develops an array of quantized vortices.

hydrogen atom. For a rotating superfluid, it leads to quantized vortices.[109] If one spins a normal liquid in a bucket, the fluid will finally rotate as a rigid body where the velocity smoothly increases from the center to the edge (Fig. 19, left). However, this smooth variation is impossible for particles in a single quantum state. To fulfill the above-mentioned quantization rule, the flow field has to develop singular regions where the number of de Broglie wavelengths on a closed path jumps up by one. One possibility would be a radially symmetric flow field with concentric rings. Between adjacent rings, the number of de Broglie wavelengths on a circumference would change by one.

However, the energetically most favorable configuration is achieved when the singularities in the velocity field are not distributed on cylindrical shells, but on lines. This corresponds to an array of vortices. In contrast to classical vortices such as in tornados or in a flushing toilet, the vortices in a Bose–Einstein condensate are quantized: when an atom goes around the vortex core, its quantum mechanical phase changes by exactly 2π. Such quantized vortices play a key role in superfluidity and superconductivity. In superconductors, magnetic flux lines arrange themselves in regular lattices that have been directly imaged. In superfluids, previous direct observations of vortices had been limited to small arrays (up to 11 vortices), both in liquid ^{4}He[110] and in rotating gaseous Bose–Einstein condensates (BEC) by a group in Paris.[111]

In 2001, our group observed the formation of highly-ordered vortex lattices in a rotating Bose-condensed gas.[112] They were produced by spinning laser beams around the condensate, thus setting it into rotation. The condensate then exhibited a remarkable manifestation of quantum mechanics at a macroscopic level. The rotating gas cloud was riddled with more than one hundred vortices. Since the vortex cores were smaller than the optical resolution, the gas was allowed to ballistically expand after the magnetic trap was switched off. This magnified the spatial structures twenty-fold. A shadow picture of these clouds showed little bright spots where the light penetrated through the empty vortex cores like trough tunnels (Fig. 20 shows a negative image).

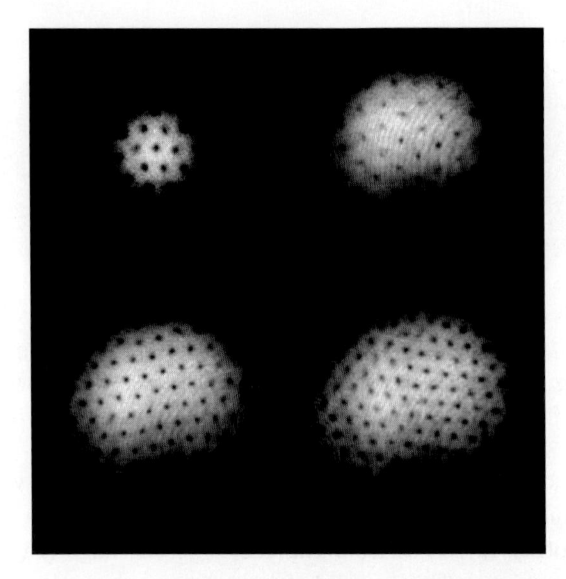

Fig. 20. Observation of vortex lattices in rotating Bose–Einstein condensates. The examples shown contain (A) 16 (B) 32 (C) 80 and (D) 130 vortices as the speed of rotation was increased. The vortices have "crystallized" in a triangular pattern. The diameter of the cloud in (D) was 1 mm after ballistic expansion, which represents a magnification of twenty. (Reprinted with permission from Ref. 112. Copyright 2001 American Association for the Advancement of Science.)

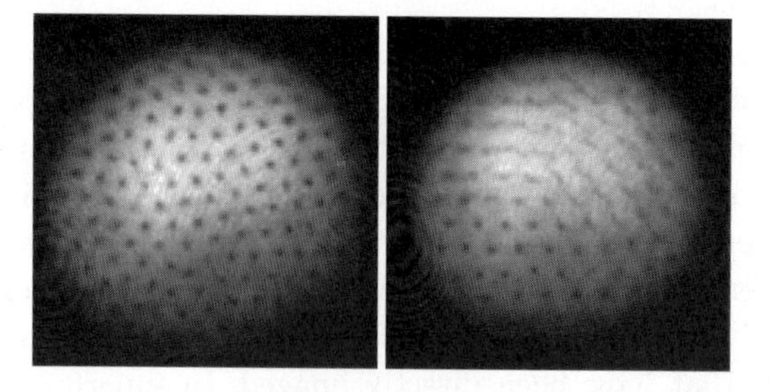

Fig. 21. Vortex lattices with defects. In the left image, the lattice has a dislocation near the center of the condensate. In the right one, there is a defect reminiscent of a grain boundary. (Reprinted with permission from Ref. 112. Copyright 2001 American Association for the Advancement of Science.)

A striking feature of the observed vortex lattices is the extreme regularity, free of any major distortions, even near the boundary. Such "Abrikosov" lattices were first predicted for quantized magnetic flux lines in type-II superconductors. However, nature is not always perfect: some of the images showed distortions or defects of the vortex lattices; two examples are shown in Fig. 21. The physics of vortices is very rich. Subsequent work by my group and others has started to address the dynamics and non-equilibrium properties of vortex structures. How are vortices formed? How do they decay? Are the vortices straight or bent? Such experiments can be directly compared with first-principle calculations, which are possible for such a dilute system. This interplay between theory and experiment may lead to a better understanding of superfluidity and macroscopic quantum phenomena.

7. Outlook

The rapid pace of developments in atomic BEC during the last few years has taken the community by surprise. After decades of an elusive search, nobody expected that condensates would be so robust and relatively easy to manipulate. Further, nobody imagined that such a simple system would pose so many challenges, not only to experimentalists, but also to our fundamental understanding of physics. The list of future challenges, both for theorists and experimentalists, is long and includes the exploration of superfluidity and second sound in Bose gases, the physics of correlations and non-classical wavefunctions (phenomena beyond the Gross–Pitaevskii equation), the study of quantum-degenerate molecules and Fermi gases, the development of practical "high-power" atom lasers, and their application in atom optics and precision measurements. These scientific goals are closely interwoven with technological advances to produce new single- or multi-species quantum-degenerate systems and novel ways of manipulation, e.g. using microtraps and atom chips. There is every indication for more excitement to come!

Work on BEC at MIT has been a tremendous team effort, and I am grateful to the past and present collaborators who have shared both the excitement

Fig. 22. The author with his team in November 2001. Front row, from left to right: Z. Hadzibabic, K. Xu, S. Gupta, E. Tsikata, Y.-I. Shin. Middle row: A. P. Chikkatur, J.-K. Chin, D. E. Pritchard, W. K., G. Campbell, A. E. Leanhardt, M. Boyd. Back row: J. R. Abo-Shaeer, D. Schneble, J. M. Vogels, K. Dieckmann, C. A. Stan, Y. Torii, E. Streed.

Fig. 23. Lecturers, seminar speakers and directors at the summer school on "Bose–Einstein Condensation in Atomic Gases" in Varenna, July 7–17, 1998. Front row: Jean Dalibard, Guglielmo Tino, Fernando Sols, Kris Helmerson. Back row: Sandro Stringari, Carl Wieman, Alexander Fetter, Tilman Esslinger, Massimo Inguscio, William Phillips, Daniel Heinzen, Peter Fedichev, Lev Pitaevskii, W. K., Allan Griffin, Keith Burnett, Daniel Kleppner, Alain Aspect, Ennio Arimondo, Theodor Hänsch, Eric Cornell.

and the hard work: J. R. Abo-Shaeer, M. R. Andrews, M. Boyd, G. Campbell, A. P. Chikkatur, J.-K. Chin, K. B. Davis, K. Dieckmann, D. S. Durfee, A. Görlitz, S. Gupta, T. L. Gustavson, Z. Hadzibabic, S. Inouye, M. A. Joffe, D. Kielpinski, M. Köhl, C. E. Kuklewicz, A. E. Leanhardt, R. F. Löw, A. Martin, M.-O. Mewes, H.-J. Miesner, R. Onofrio, T. Pfau, D. E. Pritchard, C. Raman, D. Schneble, C. Schunck, Y.-I. Shin, D. M. Stamper-Kurn, C. A. Stan, J. Stenger, E. Streed, Y. Torii, C. G. Townsend, N. J. van Druten, J. M. Vogels, K. Xu, M. W. Zwierlein, and many MIT undergraduate students. Exemplary administrative support has been provided by Carol Costa for more than twelve years. Figure 22 shows the team in November 2001. Special thanks go to Dan Kleppner and Tom Greytak for inspiration and constant encouragement. The author also acknowledges the fruitful interactions with colleagues all over the world who have contributed to this rich and exciting field. Some of these colleagues are depicted in Fig. 23, which is a group photo of the lecturers at the Varenna summer school on BEC in 1998. In particular, the year long competition with the group at Boulder led by Eric Cornell and Carl Wieman inspired the best from me and my team, and despite tight competition, there has been genuine collegiality and friendship. I want to thank the Office of Naval Research, the National Science Foundation, the Army Research Office, the Joint Services Electronics Program, NASA and the David and Lucile Packard Foundation for their encouragement and financial support of this work.

References

1. A. Einstein, *Sitzungsber. Preuss. Akad. Wiss., Bericht* 3, 18 (1925).
2. S. N. Bose, *Z. Phys.* **26**, 178 (1924).
3. G. Taubes, *Science* **265**, 184 (1994).
4. S. Chu, *Rev. Mod. Phys.* **70**, 685 (1998).
5. C. N. Cohen-Tannoudji, *Rev. Mod. Phys.* **70**, 707 (1998).
6. W. D. Phillips, *Rev. Mod. Phys.* **70**, 721 (1998).
7. N. Masuhara, J. M. Doyle, J. C. Sandberg, D. Kleppner, T. J. Greytak, H. F. Hess and G. P. Kochanski, *Phys. Rev. Lett.* **61**, 935 (1988).
8. W. Ketterle and N. J. van Druten, in *Advances in Atomic, Molecular and Optical Physics*, eds. B. Bederson and H. Walther (Academic Press, San Diego, 1996), Vol. 37, pp. 181–236.
9. J. T. M. Walraven, *Quantum Dynamics of Simple Systems*, eds. G. L. Oppo, S. M. Barnett, E. Riis and M. Wilkinson (Institute of Physics Publ., London, 1996), pp. 315–352.
10. W. Ketterle, D. S. Durfee and D. M. Stamper-Kurn, in *Bose–Einstein Condensation in Atomic Gases*, Proceedings of the International School of Physics Enrico Fermi, Course CXL, eds. M. Inguscio, S. Stringari and C. E. Wieman (IOS Press, Amsterdam, 1999), pp. 67–176.
11. D. Stamper-Kurn and W. Ketterle, in *Coherent Atomic Matter Waves*, Proceedings of the Les Houches Summer School, Course LXXII in 1999, eds. R. Kaiser, C. Westbrook and F. David (Springer, NewYork, 2001), e-print cond-mat/0005001.
12. W. Ketterle and S. Inouye, Compte Rendus de l'Académie des Sciences, Sèrie IV — Physique Astrophysique, **2**, 339–380 (2001); e-print cond-mat/0101424.
13. A. Pais, *Subtle is the Lord, The Science and the Life of Albert Einstein* (Clarendon Press, Oxford, 1982).
14. K. Gavroglu, *Fritz London: A Scientific Biography* (Cambridge University Press, Cambridge, 1995).
15. A. Griffin, in *Bose–Einstein Condensation in Atomic Gases*, Proceedings of the International School of Physics Enrico Fermi, Course CXL, eds. M. Inguscio, S. Stringari and C. E. Wieman (IOS Press, Amsterdam, 1999), p. 1.
16. A. Einstein, *Sitzungsber. Preuss. Akad. Wiss.*, Bericht 1, 3 (1925).
17. A. Griffin, D. W. Snoke and S. Stringari (editors), *Bose–Einstein Condensation* (Cambridge University Press, Cambridge, 1995).
18. K. Huang, *Statistical Mechanics* (Wiley, New York, 1987).
19. E. Schrödinger, *Statistical Thermodynamics* (Cambridge University Press, Cambridge, 1952), reprinted by Dover Pulblications (New York, 1989).
20. B. C. Crooker, B. Hebral, E. N. Smith, Y. Takano and J. D. Reppy, *Phys. Rev. Lett.* **51**, 666 (1983).
21. J. D. Reppy, *Physica* **B126**, 335 (1984).
22. M. Rasolt, M. H. Stephen, M. E. Fisher and P. B. Weichman, *Phys. Rev. Lett.* **53**, 798 (1984).
23. H. Cho and G. A. Williams, *Phys. Rev. Lett.* **75**, 1562 (1995).
24. J. P. Wolfe, J. L. Lin and D. W. Snoke, in *Bose–Einstein Condensation*, eds. A. Griffin, D. W. Snoke and S. Stringari (Cambridge University Press, Cambridge, 1995), pp. 281–329.
25. E. Fortin, E. Benson and A. Mysyrowicz, in *Bose–Einstein Condensation*, eds. A. Griffin, D. W. Snoke and S. Stringari (Cambridge University Press, Cambridge, 1995), pp. 519–523.
26. J. L. Lin and J. P. Wolfe, *Phys. Rev. Lett.* **71**, 1222 (1993).

27. K. E. O'Hara, L. Ó Súilleabháin and J. P. Wolfe, *Phys. Rev.* **B60**, 10565 (1999).
28. L. V. Butov, C. W. Lai, A. L. Ivanov, A. C. Gossard and D. S. Chemla, *Nature* **417**, 47 (2002).
29. Y. Yamamoto, *Nature* **405**, 629 (2000).
30. J. J. Baumberg, *Phys. World*, March 2002, p. 37.
31. M. Saba, C. Ciuti, J. Bloch, V. Thierry-Mieg, R. Andr'e, L. S. Dang, S. Kundermann, A. Mura, G. Bongiovanni, J. L. Staehli and B. Deveaud, *Nature* **414**, 731 (2001).
32. C. E. Hecht, *Physica* **25**, 1159 (1959).
33. W. C. Stwalley and L. H. Nosanow, *Phys. Rev. Lett.* **36**, 910 (1976).
34. I. F. Silvera and J. T. M. Walraven, *Phys. Rev. Lett.* **44**, 164 (1980).
35. R. W. Cline, D. A. Smith, T. J. Greytak and D. Kleppner, *Phys. Rev. Lett.* **45**, 2117 (1980).
36. D. G. Fried, T. C. Killian, L. Willmann, D. Landhuis, S. C. Moss, D. Kleppner and T. J. Greytak, *Phys. Rev. Lett.* **81**, 3811 (1998).
37. T. J. Greytak, in *Bose–Einstein Condensation*, eds. A. Griffin, D. W. Snoke and S. Stringari (Cambridge University Press, Cambridge, 1995), pp. 131–159.
38. T. J. Greytak and D. Kleppner, in *New Trends in Atomic Physics*, Les Houches Summer School 1982, eds. G. Grynberg and R. Stora (North-Holland, Amsterdam, 1984), p. 1125.
39. I. F. Silvera and J. T. M. Walraven, in *Progress in Low Temperature Physics*, ed. D. F. Brewer (Elsevier, Amsterdam, 1986), Vol. X, p. 139.
40. D. Kleppner, T. J. Greytak, T. C. Killian, D. G. Fried, L. Willmann, D. Landhuis and S. C. Moss, in *Bose–Einstein Condensation in Atomic Gases*, Proceedings of the International School of Physics Enrico Fermi, Course CXL, eds. M. Inguscio, S. Stringari and C. E. Wieman (IOS Press, Amsterdam, 1999), pp. 177–199.
41. A. I. Safonov, S. A. Vasilyev, I. S. Yasnikov, I. I. Lukashevich and S. Jaakola, *Phys. Rev. Lett.* **81**, 4545 (1998).
42. V. V. Goldman, I. F. Silvera and A. J. Leggett, *Phys. Rev.* **B24**, 2870 (1981).
43. D. A. Huse and E. Siggia, *J. Low Temp. Phys.* **46**, 137 (1982).
44. J. Oliva, *Phys. Rev.* **B39**, 4197 (1989).
45. H. T. C. Stoof, J. M. V. A. Koelman and B. J. Verhaar, *Phys. Rev.* **B38**, 4688 (1988).
46. H. F. Hess, *Phys. Rev.* **B34**, 3476 (1986).
47. E. Arimondo, W. D. Phillips and F. Strumia (editors), *Laser Manipulation of Atoms and Ions*, Proceedings of the International School of Physics Enrico Fermi, Course CXVIII (North-Holland, Amsterdam, 1992).
48. H. Metcalf and P. van der Straten, *Phys. Rep.* **244**, 203 (1994).
49. C. S. Adams and E. Riis, *Progress in Quantum Electronics* **21**, 1 (1997).
50. V. S. Letokhov and V. G. Minogin, *Optics Commun.* **35**, 199 (1980).
51. S. Chu, L. Hollberg, J. E. Bjorkholm, A. Cable and A. Ashkin, *Phys. Rev. Lett.* **55**, 48 (1985).
52. D. E. Pritchard, in *Electronic and Atomic Collisions*, invited papers of the XIV International Conference on the Physics of Electronic and Atomic Collisions, Palo Alto, California, 24–30 July, 1985, eds. D. C. Lorents, W. E. Meyerhof and J. R. Peterson (Elsevier, New York, 1986), pp. 593–604.
53. T. Walker and P. Feng, in *Advances in Atomic, Molecular and Optical Physics*, eds. B. Bederson and H. Walther (Academic Press, San Diego, 1994), Vol. 34, pp. 125–170.
54. J. Weiner, in *Advances in Atomic, Molecular and Optical Physics*, eds. B. Bederson and H. Walther (Academic Press, San Diego, 1995), Vol. 35, pp. 45–78.
55. T. Walker, D. Sesko and C. Wieman, *Phys. Rev. Lett.* **64**, 408 (1990).

56. M. Drewsen, P. Laurent, A. Nadir, G. Santarelli, A. Clairon, Y. Castin, D. Grison and C. Salomon, *Appl. Phys.* **B59**, 283 (1994).
57. C. G. Townsend, N. H. Edwards, C. J. Cooper, K. P. Zetie, C. J. Foot, A. M. Steane, P. Szriftgiser, H. Perrin and J. Dalibard, *Phys. Rev.* **A52**, 1423 (1995).
58. C. G. Townsend, N. H. Edwards, K. P. Zetie, C. J. Cooper, J. Rink and C. J. Foot, *Phys. Rev.* **A53**, 1702 (1996).
59. M. T. DePue, C. McCormick, S. L. Winoto, S. Oliver and D. S. Weiss, *Phys. Rev. Lett.* **82**, 2262 (1999).
60. T. Ido, Y. Isoya and H. Katori, *Phys. Rev.* **A61**, 061403(R) (2000).
61. A. J. Kerman, V. Vuletic, C. Chin and S. Chu, *Phys. Rev. Lett.* **84**, 439 (2000).
62. W. Ketterle and D. E. Pritchard, *Appl. Phys.* **B54**, 403 (1992).
63. R. V. E. Lovelace, C. Mahanian, T. J. Tommila and D. M. Lee, *Nature* **318**, 30 (1985).
64. E. A. Cornell, C. Monroe and C. E. Wieman, *Phys. Rev. Lett.* **67**, 2439 (1991).
65. J. Vigué, *Phys. Rev.* **A34**, 4476 (1986).
66. W. Ketterle and D. E. Pritchard, *Phys. Rev.* **A46**, 4051 (1992).
67. D. E. Pritchard and W. Ketterle, in *Laser Manipulation of Atoms and Ions*, Proceedings of the International School of Physics Enrico Fermi, Course CXVIII, eds. E. Arimondo, W. D. Phillips and F. Strumia (North-Holland, Amsterdam, 1992), pp. 473–496.
68. M. Kasevich and S. Chu, *Phys. Rev. Lett.* **69**, 1741 (1992).
69. C. Monroe, W. Swann, H. Robinson and C. Wieman, *Phys. Rev. Lett.* **65**, 1571 (1990).
70. K. Lindquist, M. Stephens and C. Wieman, *Phys. Rev.* **A46**, 4082 (1992).
71. W. Ketterle, A. Martin, M. A. Joffe and D. E. Pritchard, *Phys. Rev. Lett.* **69**, 2483 (1992).
72. T. E. Barrett, S. W. Dapore-Schwartz, M. D. Ray and G. P. Lafyatis, *Phys. Rev. Lett.* **67**, 3483 (1991).
73. M. A. Joffe, W. Ketterle, A. Martin and D. E. Pritchard, *J. Opt. Soc. Am.* **B10**, 2257 (1993).
74. D. E. Pritchard, K. Helmerson and A. G. Martin, in *Atomic Physics 11*, eds. S. Haroche, J. C. Gay and G. Grynberg (World Scientific, Singapore, 1989), p. 179.
75. A. L. Migdall, J. V. Prodan, W. D. Phillips, T. H. Bergeman and H. J. Metcalf, *Phys. Rev. Lett.* **54**, 2596 (1985).
76. V. S. Bagnato, G. P. Lafyatis, A. G. Martin, E. L. Raab, R. N. Ahmad-Bitar and D. E. Pritchard, *Phys. Rev. Lett.* **58**, 2194 (1987).
77. K. Helmerson, A. Martin and D. E. Pritchard, *J. Opt. Soc. Am.* **B9**, 1988 (1992).
78. C. Monroe, E. Cornell and C. Wieman, in *Laser Manipulation of Atoms and Ions*, Proceedings of the International School of Physics Enrico Fermi, Course CXVIII, eds. E. Arimondo, W. D. Phillips and F. Strumia (North-Holland, Amsterdam, 1992), pp. 361–377.
79. C. C. Bradley, C. A. Sackett, J. J. Tollet and R. G. Hulet, *Phys. Rev. Lett.* **75**, 1687 (1995).
80. W. Ketterle, K. B. Davis, M. A. Joffe, A. Martin and D. E. Pritchard, *Phys. Rev. Lett.* **70**, 2253 (1993).
81. J. Weiner, V. S. Bagnato, S. Zilio and P. S. Julienne, *Rev. Mod. Phys.* **71**, 1 (1999).
82. D. J. Heinzen, in *Bose–Einstein Condensation in Atomic Gases*, Proceedings of the International School of Physics Enrico Fermi, Course CXL, eds. M. Inguscio, S. Stringari and C. E. Wieman (IOS Press, Amsterdam, 1999), pp. 351–390.
83. M. H. Anderson, J. R. Ensher, M. R. Matthews, C. E. Wieman and E. A. Cornell, *Science* **269**, 198 (1995).

84. K. B. Davis, M.-O. Mewes, M. R. Andrews, N. J. van Druten, D. S. Durfee, D. M. Kurn and W. Ketterle, *Phys. Rev. Lett.* **75**, 3969 (1995).
85. W. Ketterle, K. B. Davis, M. A. Joffe, A. Martin and D. E. Pritchard, Presentation at the *OSA Annual Meeting* (Toronto, Canada, October 3–8, 1993).
86. P. D. Lett, W. D. Phillips, S. L. Rolston, C. E. Tanner, R. N. Watts and C. I. Westbrook, *J. Opt. Soc. Am.* **B6**, 2084 (1989).
87. K. B. Davis, M. O. Mewes, M. A. Joffe and W. Ketterle, in *14th Int. Conf. Atomic Phys.* Book of Abstracts, 1-M3 (University of Colorado, Boulder, Colorado, 1994).
88. W. Petrich, M. H. Anderson, J. R. Ensher and E. A. Cornell, *Phys. Rev. Lett.* **74**, 3352 (1995).
89. K. B. Davis, M.-O. Mewes, M. A. Joffe, M. R. Andrews and W. Ketterle, *Phys. Rev. Lett.* **74**, 5202 (1995).
90. J. M. Doyle, J. C. Sandberg, I. A. Yu, C. L. Cesar, D. Kleppner and T. J. Greytak, *Phys. Rev. Lett.* **67**, 603 (1991).
91. M. O. Mewes, M. R. Andrews, N. J. van Druten, D. M. Kurn, D. S. Durfee and W. Ketterle, *Phys. Rev. Lett.* **77**, 416 (1996).
92. D. S. Durfee and W. Ketterle, *Optics Express* **2**, 299 (1998).
93. C. C. Bradley, C. A. Sackett, J. J. Tollet and R. G. Hulet, *Phys. Rev. Lett.* **79**, 1170 (1997).
94. W. Petrich, M. H. Anderson, J. R. Ensher and E. A. Cornell, *J. Opt. Soc. Am.* **B11**, 1332 (1994).
95. E. L. Surkov, J. T. M. Walraven and G. V. Shlyapnikov, *Phys. Rev.* **D53**, 3403 (1996).
96. M. R. Andrews, M.-O. Mewes, N. J. van Druten, D. S. Durfee, D. M. Kurn and W. Ketterle, *Science* **273**, 84 (1996).
97. M.-O. Mewes, M. R. Andrews, N. J. van Druten, D. M. Kurn, D. S. Durfee, C. G. Townsend and W. Ketterle, *Phys. Rev. Lett.* **77**, 988 (1996).
98. M.-O. Mewes, M. R. Andrews, D. M. Kurn, D. S. Durfee, C. G. Townsend and W. Ketterle, *Phys. Rev. Lett.* **78**, 582 (1997).
99. C. G. Townsend, N. J. van Druten, M. R. Andrews, D. S. Durfee, D. M. Kurn, M.-O. Mewes and W. Ketterle, in *Atomic Physics 15*, Fifteenth International Conference on Atomic Physics, Amsterdam, August 1996, eds. H. B. van Linden van den Heuvell, J. T. M. Walraven and M. W. Reynolds (World Scientific, Singapore, 1997), pp. 192–211.
100. A. Röhrl, M. Naraschewski, A. Schenzle and H. Wallis, *Phys. Rev. Lett.* **78**, 4143 (1997).
101. M. R. Andrews, C. G. Townsend, H.-J. Miesner, D. S. Durfee, D. M. Kurn and W. Ketterle, *Science* **275**, 637 (1997).
102. M. Inguscio, S. Stringari and C. E. Wieman (editors), *Bose–Einstein Condensation in Atomic Gases*, Proceedings of the International School of Physics Enrico Fermi, Course CXL (IOS Press, Amsterdam, 1999).
103. C. S. Adams, M. Sigel and J. Mlynek, *Phys. Rep.* **240**, 143 (1994).
104. H.-J. Miesner, D. M. Stamper-Kurn, M. R. Andrews, D. S. Durfee, S. Inouye and W. Ketterle, *Science* **279**, 1005 (1998).
105. S. Inouye, T. Pfau, S. Gupta, A. P. Chikkatur, A. Görlitz, D. E. Pritchard and W. Ketterle, *Nature* **402**, 641 (1999).
106. M. Kozuma, Y. Suzuki, Y. Torii, T. Sugiura, T. Kuga, E. W. Hagley and L. Deng, *Science* **286**, 2309 (1999).
107. S. Inouye, A. P. Chikkatur, D. M. Stamper-Kurn, J. Stenger, D. E. Pritchard and W. Ketterle, *Science* **285**, 571 (1999).

108. J. Stenger, S. Inouye, A. P. Chikkatur, D. M. Stamper-Kurn, D. E. Pritchard and W. Ketterle, *Phys. Rev. Lett.* **82**, 4569 (1999).
109. P. Nozières and D. Pines, *The Theory of Quantum Liquids* (Addison-Wesley, Redwood City, CA, 1990).
110. E. J. Yarmchuk, M. J. V. Gordon and R. E. Packard, *Phys. Rev. Lett.* **43**, 214 (1979).
111. K. W. Madison, F. Chevy, W. Wohlleben and J. Dalibard, *J. Mod. Opt.* **47**, 2715 (2000).
112. J. R. Abo-Shaeer, C. Raman, J. M. Vogels and W. Ketterle, *Science* **292**, 476 (2001).

G. E. Brown

A THEORY OF GAMMA-RAY BURSTS*

G. E. BROWN†, C.-H. LEE‡ and R. A. M. J. WIJERS§
Department of Physics & Astronomy, State University of New York,
Stony Brook, New York 11794, USA
†*popenoe@nuclear.physics.sunysb.edu*

H. K. LEE
Department of Physics, Hanyang University, Seoul 133-791, South Korea
hklee@hepth.hanyang.ac.kr

G. ISRAELIAN
Instituto de Astrofisica de Canarias, E-38200 La Laguna, Tenerife, Spain
gil@ll.iac.es

H. A. BETHE
Floyd R. Newman Laboratory of Nuclear Studies, Cornell University,
Ithaca, New York 14853, USA

Recent observations and theoretical considerations have linked gamma-ray bursts with ultra-bright type Ibc supernovae ('hypernovae'). We here work out a specific scenario for this connection. Based on earlier work, we argue that especially the longest bursts must be powered by the Blandford–Znajek mechanism of electromagnetic extraction of spin energy from a black hole. Such a mechanism requires a high angular momentum in the progenitor object. The observed association of gamma-ray bursts with type Ibc supernovae leads us to consider massive helium stars that form black holes at the end of their lives as progenitors. In our analysis we combine the numerical work of MacFadyen and Woosley with analytic calculations in Kerr geometry, to show that about 10^{53} erg each are available to drive the fast GRB ejecta and the supernova. The GRB ejecta are driven by the power output through the open field lines threading the black hole, whereas the supernova can be powered both by the shocks driven into the envelope by the jet, and by the power delivered into the disk via field lines connecting the disk with the black hole. We also present a much simplified approximate derivation of these energetics.

Helium stars that leave massive black-hole remnants can only be made in fairly specific binary evolution scenarios, namely the kind that also leads to the formation of soft X-ray transients with black-hole primaries, or in very massive WNL stars. Since

*First published in *New Astronomy* **5** (2000) 191–210 (© 2000 Elsevier Science B.V.).
‡Current and permanent address: Department of Physics, Pusan National University, Pusan 609-735, Korea. E-mail: clee@pusan.ac.kr
§Current and permanent address: University of Amsterdam, Kruislaan 403, NL-1098 SJ Amsterdam, The Netherlands. E-mail: rwijers@astro.uva.nl

the binary progenitors will inevitably possess the high angular momentum we need, we propose a natural link between balck-hole transients and gamma-ray bursts. Recent observations of one such transient, GRO J1655-40/Nova Scorpii 1994, explicitly support this connection: its high space velocity indicates that substantial mass was ejected in the formation of the black hole, and the overabundance of α-nuclei, especially sulphur, indicates that the explosion energy was extreme, as in SN 1998bw/GRB 980425. Furthermore, X-ray studies of this object indicate that the black hole may still be spinning quite raidly, as expected in our model. We also show that the presence of a disk during the powering of the GRB and the explosion is required to deposit enough of the α nuclei on the companion.

Keywords: Accretion; accretion disks; Black hole physics; MHD; Supernovae: general; Gamma-rays: bursts; Gamma-rays: theory.

PACS: 98.70.Rz; 97.10.Cv; 97.10.Tk; 97.60.Lf; 95.30.Sf

1. Introduction

The discovery of afterglows to gamma-ray bursts has greatly increased the possibility of studying their physics. Since these afterglows have thus far only been seen for long gamma-ray bursts (duration $\gtrsim 2$ s), we shall concentrate on the mechanism for this subclass. The shorter bursts (duration $\lesssim 2$ s) may have a different origin; specifically, it has been suggested that they are the result of compact-object mergers and therefore offer the intriguing possibility of associated outbursts of gravity waves. (Traditionally, binary neutron stars have been considered in this category (Eichler *et al.*, 1989; Janka *et al.*, 1999). More recently, Bethe and Brown (1998) have shown that low-mass black-hole, neutron-star binaries, which have a ten times greater formation rate and are stronger gravity-wave emitters, may be the more promising source of this kind.)

An important recent clue to the origin of long bursts is the probable association of some of them with ultra-bright type Ibc supernovae (Bloom *et al.*, 1999; Galama *et al.*, 1998, 2000). The very large explosion energy[1] implied by fitting the light curve of SN 1998bw, which was associated with GRB 980425 (Galama *et al.*, 1998), indicates that a black hole was formed in this event (Iwamoto *et al.*, 1998). This provides two good pieces of astrophysical information: it implicates black holes in the origin of gamma-ray bursts, and it demonstrates that a massive star can explode as a supernova even if its core collapses into a black hole.

In this paper, we start from the viewpoint that the gamma-ray burst is powered by electromagnetic energy extraction from a spinning black hole, the so-called Blandford–Znajek mechanism (Blandford and Znajek, 1977). This was worked out in detail by Lee *et al.* (1999), and further details and comments were discussed by Lee *et al.* (2000), who built on work by Thorne *et al.* (1986) and Li (2000). They have shown that with the circuitry in a $3+1$ dimensional description

[1]Höflich *et al.* (1999) have proposed that the explosion energy was not much larger than usual, but that the explosion was very asymmetric; this model also provides a reasonable fit to the light curve of SN 1998bw.

using the Boyer-Lindquist metric, one can have a simple pictorial model for the BZ mechanism.

The simple circuitry which involves steady state current flow is, however, inadequate for describing dissipation of the black hole rotational energy into the accretion disk formed from the original helium envelope. In this case the more rapidly rotating black hole tries to spin up the inner accretion disk through the closed field lines coupling the black hole and disk. Electric and magnetic fields vary wildly with time. Using the work of Blandford and Spruit (2000) we show that this this dissipation occurs in an oscillatory fashion, giving a fine structure to the GRB, and that the total dissipation should furnish an energy comparable to that of the GRB to the accretion disk. We use this energy to drive the hypernova explosion.

Not any black-hole system will be suitable for making GRB: the black hole must spin rapidly enough and be embedded in a strong magnetic field. Moreover, the formation rate must be high enough to get the right rate of GRB even after accounting for substantial collimation of GRB outflows. We explore a variety of models, and give arguments why some will have sufficient energy and extraction efficiency to power a GRB and a hypernova. We argue that the systems known as black-hole transients are the relics of GRBs, and discuss the recent evidence from high space velocities and chemical abundance anomalies that these objects are relics of hypernovae and GRBs; we especially highlight the case of Nova Scorpii 1994 (GRO J1655-40).

The plan of this paper is as follows. We first show that it is reasonable to expect similar energy depositions into the GRB outflow and the accretion disk (Sec. 2) and discuss the amount of available energy to be extracted (Sec. 3). Then we show the agreement of those results with the detailed numerical simulations by MacFadyen and Woosley, and use those simulations to firm up our numbers (Sec. 4). We continue by presenting a simple derivation of the energetics that approximates the full results well (Sec. 5). Finally, we discuss some previously suggested progenitors (Sec. 6) and present our preferred progenitors: soft X-ray transients (Sec. 7).

2. Simple Circuitry

Although our numbers are based on the detailed review of Lee *et al.* (1999), which confirms the original Blandford–Znajek paper (Blandford and Znajek, 1977), we illustrate our arguments with the pictorial treatment of Thorne *et al.* (1986) in "*The Membrane Paradigm*". Considering the time as universal in the Boyer-Lindquist metric, essential electromagnetic and statistical mechanics relations apply in their $3+1$ dimensional manifold. We summarize their picture in our Fig. 1.

The surface of the black hole can be considered as a conductor with surface resistance $R_{\rm BH} = 4\pi/c = 377\ \Omega$. A circuit that rotates rigidly with the black hole can be drawn from the loading region, the low-field region up the axis of rotation of the black hole in which the power to run the GRB is delivered, down a magnetic field line, then from the North pole of the black hole along the (stretched) horizon to

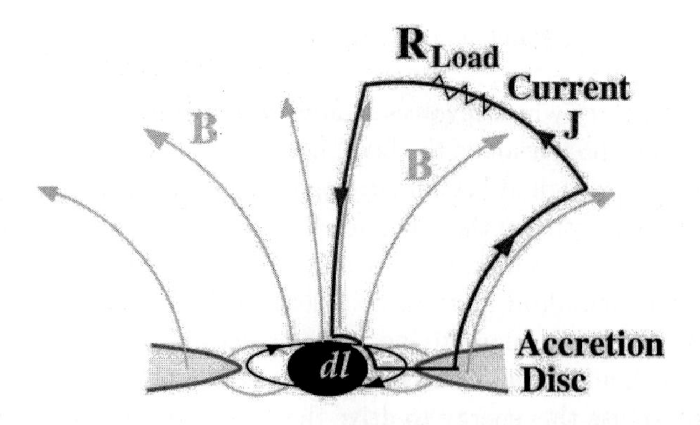

Fig. 1. The black hole in rotation about the accretion disk. A circuit, in rigid rotation with the black hole is shown. This circuit cuts the field lines from the disk as the black hole rotates, and by Faraday's law, produces an electromotive force. This force drives a current. More detailed discussion is given in the text.

its equator. From the equator we continue the circuit through part of the disk and then connect it upwards with the loading region. We can also draw circuits starting from the loading region which pass along only the black hole or go through only the disk, but adding these would not change the results of our schematic model.

Using Faraday's law, the voltage V can be found by integrating the vector product of charge velocity, v, and magnetic field, $\boldsymbol{B}$, along the circuit:

$$V = \int [\boldsymbol{v} \times \boldsymbol{B}] \cdot \boldsymbol{dl} \,, \tag{1}$$

($\boldsymbol{dl}$ is the line element along the circuit). Because this law involves $\boldsymbol{v} \times \boldsymbol{B}$ the integrals along the field lines make no contribution. We do get a contribution V from the integral from North pole to equator along the black hole surface. Further contributions to V will come from cutting the field lines from the disk. We assume the field to be weak enough in the loading region to be neglected.

The GRB power, $E_{\rm GRB}$, will be

$$\dot{E}_{\rm GRB} = I^2_{\rm BH+D} R_{\rm L} \,, \tag{2}$$

where $R_{\rm L}$ is the resistance of the loading region, and the current is given by

$$I^2_{\rm BH+D} = \left(\frac{V_{\rm D} + V_{\rm BH}}{R_{\rm D} + R_{\rm BH} + R_{\rm L}} \right)^2 . \tag{3}$$

(The index BH refers to the black hole, L to the load region, and D to the disk.)

The load resistance has been estimated in various ways and for various assumptions by Lovelace *et al.* (1979) and by MacDonald and Thorne (1982), and by Phinney (1983). All estimates agree that to within a factor of order unity $R_{\rm L}$ is equal to $R_{\rm BH}$.

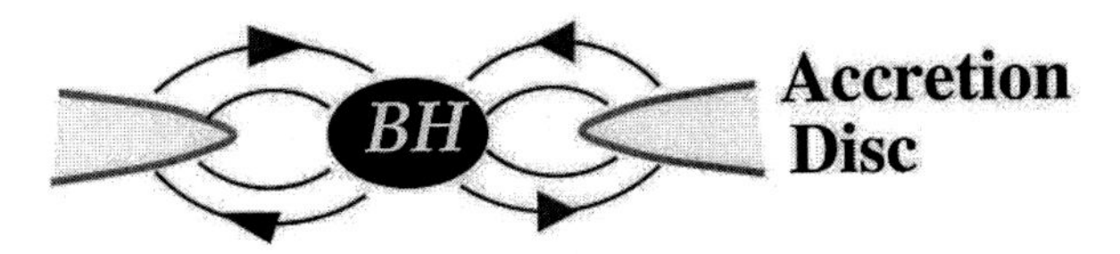

Fig. 2. Magnetic field lines, anchored in the disk, which thread the black hole, coupling the disk rotation to that of the black hole.

In a similar fashion, some power will be deposited into the disk

$$\dot{E}_{\text{disk}} = I^2_{\text{BH+D}} R_{\text{D}} \tag{4}$$

but this equilibrium contribution will be small because of the low disk resistance R_{D}.

Blandford and Spruit (2000) have shown that important dissipation into the disk comes through magnetic field lines coupling the disk to the black hole rotation. As shown in Fig. 2 these lines, anchored in the inner disk, thread the black hole.

The more rapidly rotating black hole will provide torques, along its rotation axis, which spin up the inner accretion disk, in which the closed magnetic field lines are anchored. With increasing centrifugal force the material in the inner disk will move outwards, cutting down the accretion. Angular momentum is then advected outwards, so that the matter can drift back inwards. It then delivers more matter to the balck hole and is flung outwards again. The situation is like that of a ball in a roulette wheel (R. D. Blandford, private communication). First of all it is flung outwards and then drifts slowly inwards. When it hits the hub it is again thrown outwards. The viscous inflow time for the fluctuations is easily estimated to be

$$\tau_{\text{d}} \sim \Omega^{-1}_{\text{disk}} \left(\frac{r}{H}\right)^2 \alpha^{-1}_{\text{vis}}, \tag{5}$$

where H is the height of the disk at radius r, Ω_{disk} its angular velocity, and α_{vis} is the usual α-parametrization of the viscosity. We choose $\alpha_{\text{vis}} \sim 0.1$, $r/H \sim 10$ for a thin disk and then arrive at $\tau_{\text{d}} \sim 0.1$ s. We therefore expect variability on all time scales between the Kepler time (sub-millisecond) and the viscous time, which may explain the very erratic light curves of many GRBs.

We suggest that the GRB can be powered by $\dot{E}_{\text{GRB}}$ and a Type Ibc supernova explosion by $\dot{E}_{\text{SN}}$ where $\dot{E}_{\text{SN}}$ is the power delivered through dissipation into the disk. To the extent that the number of closed field lines coupling disk and black hole is equal to the number of open field lines threading the latter, the two energies will be equal. In the spectacular case of GRB 980326 (Bloom *et al.*, 1999), the GRB lasts about 5 s, which we take to be the time that the central engine operates. We shall show that up to $\sim 10^{53}$ erg is available to be delivered into the GRB and into the accretion disk, the latter helping to power the supernova (SN) explosion. This is more energy than needed and we suggest that injection of energy into the disk shuts off the central engine by blowing up the disk and thus removing the magnetic field needed for the energy extraction from the black hole. If the magnetic field is high enough the energy will be delivered in a short time, and the quick removal of the disk will leave the black hole still spinning quite rapidly.

3. Energetics of GRBs

The maximum energy that can be extracted from the BZ mechanism (Lee *et al.*, 1999) is

$$(E_{\rm BZ})_{\rm max} \simeq 0.09 M_{\rm BH} c^2 . \tag{6}$$

This is 31% of the black hole rotational energy, the remainder going toward increasing the entropy of the black hole. This maximum energy is obtained if the extraction efficiency is

$$\epsilon_\Omega = \frac{\Omega_{\rm disk}}{\Omega_{\rm H}} = 0.5 . \tag{7}$$

In Appendix A we give numerical estimates for this ratio for various $\omega = \Omega_{\rm disk}/\Omega_{\rm K}$ and various radii in the region of parameter space we consider. As explained in Sec. 2 we expect the material in the inner disk to swing in and out around the marginally stable radius, $r_{\rm ms}$. It can be seen from the Table 2 and Appendix A that the relevant values of ϵ_Ω are close to that of Eq. (7).

For a $7 M_\odot$ black hole, such as that found in Nova Sco 1994 (GRO J1655-40),

$$E_{\rm max} \simeq 1.1 \times 10^{54} \text{ erg}. \tag{8}$$

We estimate below that the energy available in a typical case will be an order of magnitude less than this. Without collimation, the estimated gamma-ray energy in GRB 990123 is about 4.5×10^{54} erg (Andersen *et al.*, 1999). The BZ scenario entails substantial beaming, so this energy should may be multiplied by $d\Omega/4\pi$, which may be a small factor (perhaps 10^{-2}).

The BZ power can be delivered at a maximum rate of

$$P_{\rm BZ} = 6.7 \times 10^{50} \left(\frac{B}{10^{15}\text{ G}}\right)^2 \left(\frac{M_{\rm BH}}{M_\odot}\right)^2 \text{erg s}^{-1} , \tag{9}$$

(Lee *et al.*, 1999) so that high magnetic fields are necessary for rapid delivery.

The above concerns the maximum energy output into the jet and the disk. The real energy available in black-hole spin in any given case, and the efficiency with which it can be extracted, depend on the rotation frequency of the newly formed black hole and the disk or torus around it. The state of the accretion disk around the newly formed black hole, and the angular momentum of the black hole, are somewhat uncertain. However, the conditions should be bracketed between a purely Keplerian, thin disk (if neutrino cooling is efficient) and a thick, noncooling hypercritical advection-dominated accretion disk (HADAF), of which we have a model (Brown *et al.*, 2000b). Let us examine the result for the Keplerian case. In terms of

$$\tilde{a} \equiv \frac{Jc}{M^2 G} , \tag{10}$$

where J is the angular momentum of the black hole, we find the rotational energy of a black hole to be

$$E_{\rm rot} = f(\tilde{a}) M c^2 , \tag{11}$$

where

$$f(\tilde{a}) = 1 - \sqrt{\frac{1}{2}(1 + \sqrt{1 - \tilde{a}^2})} . \tag{12}$$

For a maximally rotating black hole one has $\tilde{a} = 1$.[2]

We begin with a neutron star in the middle of a Keplerian accretion disk, and let it accrete enough matter to send it into a black hole. In matter free regions the last stable orbit of a particle around a black hole in Schwarzschild geometry is

$$r_{\rm lso} = 3R_{\rm Sch} = 6\frac{GM}{c^2} . \tag{13}$$

This is the marginally stable orbit $r_{\rm ms}$. However, under conditions of hypercritical accretion, the pressure and energy profiles are changed and it is better to use (Abramowicz *et al.*, 1988)

$$r_{\rm lso} \gtrsim 2R_{\rm Sch} . \tag{14}$$

With the equal sign we have the marginally bound orbit $r_{\rm mb}$. With high rates of accretion we expect this to be a good approximation to $r_{\rm lso}$. The accretion disk can be taken to extend down to the last stable orbit (refer to Appendix B for the details).

We take the angular velocity to be Keplerian, so that the disk velocity v at radius $2R_{\rm Sch}$ is given by

$$v^2 = \frac{GM}{2R_{\rm Sch}} = \frac{c^2}{4} , \tag{15}$$

or $v = c/2$. The specific angular momentum, l, is then

$$l \geq 2R_{\rm Sch} v = 2\frac{GM}{c} , \tag{16}$$

which in Kerr geometry indicates $\tilde{a} \sim 1$. Had we taken one of the slowest-rotating disk flows that are possible, the advection-dominated or HADAF case (Brown *et al.*, 2000b; Narayan and Yi, 1994), which has $\Omega^2 = 2\Omega_{\rm K}^2/7$, we would have arrived at $\tilde{a} \sim 0.54$, so the Kerr parameter will always be high.

Further acceretion will add angular momentum to the black hole at a rate determined by the angular velocity of the inner disk. The material accreting into

[2] As an aside, we note a nice mnemonic: if we define a velocity v from the black-hole angular momentum by $J = MR_{\rm Sch}v$, so that v carries the quasi-interpretation of a rotation velocity at the horizon, then $\tilde{a} = 2v/c$. A maximal Kerr hole, which has $R_{\rm event} = R_{\rm Sch}/2$, thus has $v = c$. For $\tilde{a} \lesssim 0.5$, the rotation energy is well approximated by the easy-to-remember expression $E_{\rm rot} = \frac{1}{2}Mv^2$.

the black hole is released by the disk at $r_{\rm lso}$, where the angular momentum delivered to the black hole is determined. This angular momentum is, however, delivered into the black hole at the event horizon $R_{\rm Sch}$, with velocity at least double that at which it is released by the disk, since the lever arm at the event horizon, is only half of that at $R_{\rm Sch}$, and angular momentum is conserved. With more rapid rotation involving movement towards a Kerr geometry where the event horizon and last stable orbit coincide at

$$r_{\rm lso} = R_{\rm event} = \frac{GM}{c^2} \,. \tag{17}$$

Although we must switch over to a Kerr geometry for quantitative results, we see that $\tilde{a}$ will not be far from its maximum value of unity. Again, for the lower angular-momentum case of a HADAF, the expected black-hole spin is not much less.

4. Comparison with Numerical Calculation

Our schematic model has the advantage over numerical calculations that one can see analytically how the scenario changes with change in parameters or assumptions. However, our model is useful only if it reproduces faithfully the results of more complete calculations which involve other effects and much more detail than we include. We here make comparison with Fig. 19 of MacFadyen and Woosley (1999). Accretion rates, etc., can be read off from their figure which we reproduce as our Fig. 3. MacFadyen and Woosley prefer $\tilde{a}_{\rm initial} = 0.5$ (We have removed their curve for $\tilde{a}_{\rm initial} = 0$). This is a reasonable value if the black hole forms from a contracting proto-neutron star near breakup. MacFadyen and Woosley find that $\tilde{a}_{\rm initial} = 0.5$ is more consistent with the angular momentum assumed for the mantle than $\tilde{a}_{\rm initial} = 0$. (They take the initial black hole to have mass 2 $M_\odot$; we choose the Brown and Bethe (1994) mass of $1.5 M_\odot$.) We confirm this in the next section.

After 5 seconds (the duration of GRB 980326) the MacFadyen and Woosley black hole mass ~ 3.2 $M_\odot$ and their Kerr parameter $\tilde{a} \sim 0.8$, which gives $f(\tilde{a})$ of our Eq. (12) of 0.11. With these parameters we find $E = 2 \times 10^{53}$ erg, available for the GRB and SN explosion.

One can imagine that continuation of the MacFadyen and Woosley curve for $M_{\rm BH}(M_\odot)$ would ultimately give something like our ~ 7 $M_\odot$, but the final black hole mass may not be relevant for our considerations. This is because more than enough energy is available to power the supernova in the first 5 seconds; as the is disrupted, the magnetic fields supported by it will also disappear, which turns off the Blandford–Znajek mechanism.

Power is delivered at the rate given by Eq. (9). Taking a black hole mass relevant here, ~ 3.2 $M_\odot$, we require a field strength of $\sim 5.8 \times 10^{15}$ G in order for our estimated energy (4×10^{52} erg) to be delivered in 5 s (the duration of GRB 980326). For such a relatively short burst, we see that the required field is quite large, but it is still not excessive if we bear in mind that magnetic fields of $\sim 10^{15}$ G have already been observed in magnetars (Kouveliotou *et al.*, 1998, 1999). Since in our

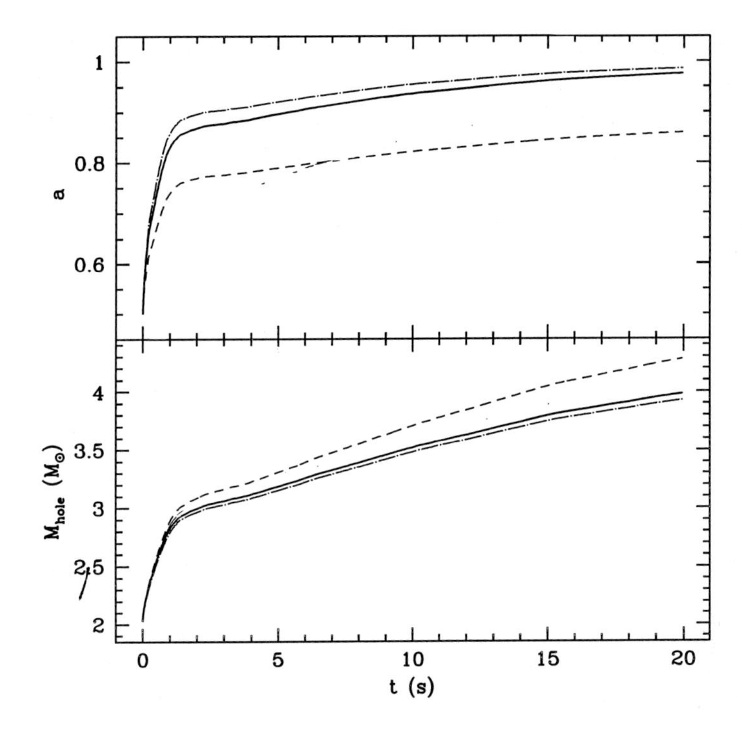

Fig. 3. Time evolution of BH mass and angular momentum taken from Fig. 19 of MacFadyen and Woosley (1999). The upper panel shows the increase in the Kerr parameter for various models for the disk interior to the inner boundary at 50 km. "Thin" (dash-dot), neutrino-dominated (thick solid) and advection dominated (short dash) models are shown for initial Kerr parameter $\tilde{a}_{\rm init} = 0.5$. The lower panel shows the growth of the gravitational mass of the black hole. The short-dashed line shows the growth in baryonic mass of the black hole since for a pure advective model no energy escapes the inner disk.

scenario we have many more progenitors than there are GRBs, we suggest that the necessary fields are obtained only in a fraction of all potential progenitors.

Thus we have an extremely simple scenario for powering a GRB and the concomitant SN explosion in the black hole transients, which we will discuss in Sec. 7.2. After the first second the newly evolved black hole has $\sim 10^{53}$ erg of rotational energy available to power these. The time scale for delivery of this energy depends (inversely quadratically) on the magnitude of the magnetic field in the neighborhood of the black hole, essentially that on the inner accretion disk. The developing supernova explosion disrupts the accretion disk; this removes the magnetic fields anchored in the disk, and selflimits the energy the B–Z mechanism can deliver.

5. An Even More Schematic Model

Here we calculate the energy available in a rotating black hole just after its birth (before accretion adds more). Our model is to take a 1.5 $M_\odot$ neutron star which co-rotates with the inner edge of the accretion disk in which it is embedded. The neutron star then collapses to a black hole, conserving its angular momentum. Since the accretion disk is neutrino cooled, but perhaps not fully thin, its angular velocity will be somewhere between the HADAF value and the Keplerian value. We parameterize it as $\Omega = \omega\Omega_K$, where $\omega = 1$ for Keplerian and $\omega = \sqrt{2/7} \sim 0.53$ for the HADAF.

The moment of inertia, I, of a neutron star is well fitted for many different equations of state with the simple expression

$$I = \frac{0.21MR^2}{1 - 2GM/Rc^2} \tag{18}$$

(Lattimer and Prakash, 2000). With $J = \omega I\Omega_K$ and a neutron star of 1.5 $M_\odot$, with a radius of 10 km, we find

$$\tilde{a}^2 = \left(\frac{Jc}{GM^2}\right)^2 = 0.64\omega^2 . \tag{19}$$

We choose $\omega \simeq 1.0$ to roughly reproduce the MacFadyen and Woosley value of $\tilde{a}$, see our Fig. 3. We do not really believe the disk to be so efficiently neutrino cooled that its angular velocity is Keplerian; i.e. $\omega = 1$, but it may be not far from it. Our ω should be more properly viewed as a fudge factor which allows us to match the more complete MacFadyen and Woosley calculation. MacFadyen and Woosley find that, while the accretion disk onto the black hole is forming, an additional solar mass of material is added to it "as the dense stellar core collapses through the inner boundary at all polar angles". We shall add this to our 1.5 $M_\odot$ and take the black hole mass to be 2.5 $M_\odot$. We neglect the increase in spin of the black hole by the newly accreted matter; this is already included in the MacFadyen and Woosley results. For $\tilde{a}^2 = 0.64$ we find $f(\tilde{a}^2) = 0.11$, so that the black hole rotation energy becomes

$$E_{BZ} = 1.5 \times 10^{53} \text{ erg} \tag{20}$$

in rough agreement with the estimates of MacFadyen and Woosley in the last section.

6. Previous Models

6.1. *Collapsar*

We have not discussed the Collapsar model of Woosley (1993), and MacFadyen and Woosley (1999). In this model the center of a rotating Wolf–Rayet star evoloves into a black hole, the outer part being held out by centrifugal force. The latter evolves into an accretion disk and then by hypercritical accretion spins the black hole up. MacFadyen and Woosley point out that "If the helium core is braked by a magnetic field prior to the supernova explosion to the extent described by Spruit and Phinney (1998) then our model will not work for single stars." Spruit and Phinney argue that magnetic fields maintained by differential rotation between the core and envelope of the star will keep the whole star in a state of approximately uniform rotation until 10 years before its collapse. As noted in the last section, with the extremely high magnetic fields we need the viscosity would be expected to be exceptionally high, making the Spruit and Phinney scenario probable. Livio and Pringle (1998) have commented that one finds evidence in novae that the coupling between layers of the star by magnetic fields may be greatly suppressed relative to what Spruit and Phinney assumed. However, we note that even with this suppressed coupling, they find pulsar periods from core collapse supernovae no shorter than 0.1 s. Independent evidence for the fact that stellar cores mostly rotate no faster than this comes from the study of supernova remnants: Bhattacharya (1990, 1991) concludes that the absence of bright, pulsar-powered plerions in most SNRs indicates that typically pulsar spin periods at birth are no shorter than 0.03–0.05 s. Translated to our black holes, such spin periods would imply $\tilde{a} \lesssim 0.01$, quite insufficient to power a GRB. As a cautionary note, we might add that without magnetic coupling the cores of evolved stars can spin quite rapidly (Heger *et al.*, 2000). This rapid initial spin many be reconciled with Bhattacharya's limit if r-mode instabilities cause very rapid spin-down in the first few years of the life of a neutronstar (e.g., Heger *et al.*, 2000; Lindblom and Owen, 1999).

6.2. *Coalescing low-mass black holes and helium stars*

Fryer and Woosley (1998) suggested the scenario of a black hole spiraling into a helium star. This is an efficient way to spin up the black hole.

Bethe and Brown (1998) evolved low-mass black holes with helium star companion, as well as binaries of compact objects. In a total available range of binary separation $0.04 < a_{13} < 4$, low-mass black-hole, neutron-star binaries were formed when $0.5 < a_{13} < 1.4$ where a_{13} is the initial binary separation in units of 10^{13} cm. The low-mass black hole coalesces with the helium star in the range $0.04 < a_{13} < 0.5$. Binaries were distributed logarithmically in a. Thus, coalescences are more common than low-mass black-hole, neutron-star binaries by a factor of $\ln(0.5/0.4)/\ln(1.9/0.5) = 1.9$.

In (Bethe and Brown, 1998), the He-star, compact-object binary was disrupted ~50% of the time by the He-star explosion. This does not apply to the coalescence. Thus, the rate of low-mass black-hole, He-star mergers is 3.8 times the formation rate of low-mass black-hole, neutron-star binaries, or

$$R = 3.8 \times 10^{-4} \text{ yr}^{-1} \tag{21}$$

in the Galaxy. The stimated empirical rate of GRBs, with a factor of 100 for beaming, is 10^{-5} yr^{-1} in the Galaxy (Appendix C of Brown *et al.* (1999a)). Thus, the number of progenitors is more than adequate.

In (Bethe and Brown, 1998) the typical black hole mass was ~2.4 $M_\odot$, somewhat more massive than their maximum assumed neutron star mass of 1.5 $M_\odot$. As it enters the helium star companion an accretion disk is soon set up and the accretion scenario will follow that described above, with rotating black holes of various masses formed. Brown *et al.* (2000b) find that the black hole will be spun up quickly. We have not pursued this scenario beyond the point that it was developed by Fryer and Woosley (1998).

7. Soft X-ray Transients as Relics of Hypernovae and GRB

7.1. *Our model: angular momentum*

We favor a model of hypernovae similar to MacFadyen and Woosley (1999) in that it involves a failed supernova as a centerpiece. But, in distinction to MacFadyen and Woosley, our initial system is a binary, consisting of a massive star A (which will later become the failed SN) and a lighter companion B, which serves to provide ample angular momentum.

Failed superovae require a ZAMS mass of 20–35 $M_\odot$, according to the calculations of Woosley and Weaver (1995) as interpreted by Brown *et al.* (1999b). The limits 20 and 35 $M_\odot$ are not accurately known, but it is a fairly narrow range, so we shall in many of our calculations assume a "typical" ZAMS mass of 25 $M_\odot$. The heavy star A must not be in a *close* binary because then its hydrogen envelope would be removed early in its evolution and there-fore the star would lose mass by wind at a very early stage and become a low-mass compact object (Brown *et al.*, 1996). Instead, we assume a wide binary, with a separation, a in the range

$$a = 500 - 1000 \ R_\odot \,, \tag{22}$$

so star A evolves essentially as a single star through its first few burning stages. It is essential that most of the He core burning is completed before its hydrogen envelope is removed (Heger and Wellstein, 2000; Wellstein and Langer, 1999). We assume the initial distance a between the two stars to be this range. When star A fills its Roche lobe, the companion, star B, will spiral inwards.

The initiation and early development of the common envelope has been best treated by Rasio and Livio (1996). This is the only phase that can at present be modeled in a realistic way. They find a short viscous time in the envelope, but emphasize that numerical viscosity may play an important role in their results. However, we believe the viscosity to be large. Torkelsson *et al.* (1996) showed the Shakura–Sunyaev viscosity parameter (Shakura and Sunyaev, 1973), α_{SS}, to range from 0.001 to 0.7, with the higher values following from the presence of vertical magnetic fields. Since in our Blandford–Znajek model extremely high magnetic fields $\sim 10^{15}$ G are needed in the He envelope to deliver the energy rapidly, we believe α_{SS} to be not much less than unity. Given such high viscosities, it seems reasonable to follow the Rasio–Livio extrapolation, based on a short viscous transport time, to later times. The most significant new result of Rasio and Livio "is that, during the dynamical phase of common envelope evolution, a corotating region of gas is established near the central binary. The corotating region has the shape of an oblate spheroid encasing the binary (i.e., the corotating gas is concentrated in the orbital plane)."

A helium core, which we deal with, is not included in their calculations, because they do not resolve the inner part of the star numerically. However, since the physics of the spiral-in does not really change as it proceeds past the end of their calculations, it seems most likely that during further spiral-in, the spin-up of material inside the orbit of the companion will continue to be significant.

Star B will stop spiraling in when it has ejected the H envelope of A. Since we assume that all stars A have about the same mass, and that a_i is very large, we expect

$$\frac{M_{\rm B}}{a_{\rm f}} \simeq \text{cont.} \tag{23}$$

From Sec. 7.2 we conclude that $a_{\rm f}$ is a few $R_\odot$ for $M_{\rm B} = (0.4-1)\ M_\odot$. Now the He cores of stars of ZAMS mass $M = 20-35 M_\odot$ have a radius about equal to $R_\odot$. Therefore small $M_{\rm B}$ stars will spiral into the He core of A. There they cannot be stopped but will coalesce with star A. However, they will have transmitted their angular momentum to star A.

Star B of larger mass will stop at larger $a_{\rm f} \gg R_\odot$. It is then not clear whether they will transfer all of their angular momentum to star A. In any case, they must generally wait until they evolve off the main sequence into the subgiant or possibly even the giant stage before they can fill their Roche Lobes and later accrete onto the black hole resulting from star A.

The Kepler velocity of star B at $a_{\rm f}$ is

$$V_{\rm K}^2 = G\frac{M_{\rm A,f}}{a_{\rm f}}\,. \tag{24}$$

We estimate the final mass of A, after removal of its hydrogen envelope, to be about 10 $M_\odot$; then

$$V_{\rm K} \simeq 1.2\times 10^8 a_{\rm f,11}^{-1/2}\ \text{cm s}^{-1}\,, \tag{25}$$

where $a_{\rm f,11}$ is $a_{\rm f}$ in units of 10^{11} cm. The specific angular momentum of B is then

$$j(B) = a_{\rm f} B_{\rm K} = 1.2\times 10^{19} a_{\rm f,11}^{1/2}\ \text{cm s}^{-1}\,. \tag{26}$$

If B and A share their angular momentum, the specific angular momentum is reduced by a factor $M_{\rm B}/(M_{\rm A,f}+M_{\rm B})$ which we estimate to be ~ 0.1. Since $a_{\rm f}$ should be $\gtrsim 3\ R_\odot$. (See Table 1), the specific angular momentum of A should be

$$j(A) \gtrsim 10^{18}\ \text{cm}^2\ \text{s}^{-1}\,. \tag{27}$$

Star B has now done its job and can be disregarded.

Table 1. Properties of transient X-ray sources.

Spectral type	$M_{\rm B}$ [$M_\odot$]	a [$R_\odot$]	$R_{\rm L}$ [$R_\odot$]	$j_{\rm B,orb}$ [10^{19} cm^2 s^{-1}]
5 K-type	0.4–0.9	4.5 ± 0.8	0.9 ± 0.2	1.5 ± 0.6
1 K-type	0.8	34	6.1	5.7
1 M-type	0.5	3.1	0.6	1.5
F (Nova Scorpii 1994)	2.2	16	4.7	.1.9
A2 (1543-47)	2.0	9.0	2.6	1.4

7.2. *Supernova and collapse*

Star A now goes through its normal evolution, ending up as a supernova. But since we have chosen its mass to be between 20 and 35 $M_\odot$, the SN shock cannot penetrate the heavy envelope but is stopped at some radius

$$R_{\rm SN} \simeq 10^{10} \text{ cm}, \tag{28}$$

well inside the outer edge of the He envelope. We estimate $R_{\rm SN}$ by scaling from SN 1987A: in that supernova, with progenitor mass $\sim$ 18 $M_\odot$, most of the He envelope was returned to the galaxy. The separation between compact object and ejecta was estimated to occur at $R \sim 5 \times 10^8$ cm (Bethe, 1990; Woosley, 1988) at mass point 1.5 $M_\odot$ (gravitational). Woosley and Weaver (1995) find remnant masses of ~ 2 $M_\odot$, although with large fluctuations, for ZAMS masses in the range 20–35 $M_\odot$, which go into high-mass black holes. From Table 3 of Brown *et al.* (1996) we see that fallback between $R = 3.5$ and 4.5×10^8 cm is 0.3 $M_\odot$. Using this we can extrapolate to $R = 10^{10}$ cm as the distance within which matter has to begin falling in immediately in our heavier stars, to make up a compact object of 2 $M_\odot$. Unlike in 1987A the shock energy in the more massive star does not suffice to eject the envelope beyond this point, and the remaining outer envelope will also eventually fall back.

At $R_{\rm SN}$, the specific angular momentum of Kepler motion around a central star of mass 10 $M_\odot$ is, cf. Eq. (26)

$$\begin{aligned} j_{\rm K}(10\ M_\odot) &= 1.2 \times 10^{19} R_{\rm f,11}^{1/2} \text{ cm}^2 \text{ s}^{-1} \\ &= 4 \times 10^{18} \text{ cm}^2 \text{ s}^{-1}\,. \end{aligned} \tag{29}$$

In reality, at this time the central object has a mass $M \sim 1.5$ $M_\odot$ (being a neutron star) and since $j_{\rm K} \sim V_{\rm K} \sim M^{1/2}$

$$j_{\rm K}(1.5\ M_\odot) = 1.5 \times 10^{18} \text{ cm}^2 \text{ s}^{-1}\,. \tag{30}$$

The therefore momentum inherent in star A, Eq. (27), is therefore greater than the Kepler angular momentum. This would not be the case had our initial object been a single star, a collapsar. (The collapsar may work none the less, but our binary model is more certain to work.)

The supernova material is supported by pressure inside the cavity, probably mostly due to electro-magnetic radiation. The cavity inside $R_{\rm SN}$ is rather free of matter. After a while, the pressure in the cavity will reduce. This may happen by opening toward the poles, in which case the outflowing pressure will drive out the matter near the poles and create the vacuum required for the gamma ray burst. Reduction of pressure will also happen by neutrino emission. As the pressure gets reduced, the SN material will fall in toward the neutron star in the center. But because the angular momentum of the SN material is large (Eq. (27)) the material must move more or less in Kepler orbits; i.e., it must spiral in. This is an essential point in the theory.

If $j(A)$ is less than $j_{\rm K}$ at $R_{\rm SN}$, the initial motion will have a substantial radial component in addition to the tangential one. But as the Kepler one decreases, cf. Eq. (29), there will come a point of r at which $j_{\rm K} = j(A)$. At this point an accretion disk will form, consisting of SN material spiraling in toward the neutron star. The primary motion is circular, but viscosity will provide a radial component inward

$$v_r \sim \alpha v_{\rm K}\,, \tag{31}$$

where α is the viscosity parameter. It has been argued by Brandenburg *et al.* (1996) that $\alpha \sim 0.1$ in the presence of equipartition magnetic fields perpendicular to the disk, and it may be even larger with the high magnetic fields required for GRBs. Narayan and Yi (1994) have given analytical solutions for such accretion disks. The material will arrive at the neutron star essentially tangentially, and therefore its high angular momentum will spin up neutron star substantially. Accretion will soon make the neutron star collapse into a black hole. The angular momentum will be conserved, so the angular velocity is increased since the black hole has smaller radius than the neutron star. Thus the black hole is born with considerable spin.

A large fraction of the material of the failed supernova will accrete onto the black hole, giving it a mass of order 7 $\rm M_\odot$. All this material adds to the angular momentum of the black hole since all of it has the Kepler velocity at the black hole radius. Our estimates show that the black hole would be close to an extreme Kerr hole (Sec. 5), were it to accrete all of this material. It may, however, be so energetic that it drives off part of the envelope in the explosion before it can all accrete (see Sec. 5).

7.3. *Soft X-ray transients with black-hole primaries*

Nine binaries have been observed which are black-hole X-ray transients. All contain a high-mass black hole, of mass $\sim$7 $\rm M_\odot$. In seven cases the lower-mass companion (star B) has a mass $\lesssim M_\odot$. The two stars are close together, their distance being of order 5 $\rm R_\odot$. Star B fills its Roche Lobe, so it spills over some material onto the black hole. The accretion disk near the black hole emits soft X rays. Two of the companions are subgiants, filling their Roche lobes at a few times larger separations from the black hole.

In fact, however, the accretion onto the central object is not constant, so there is usually no X-ray emission. Instead, the material forms an accretion disk around the black hole, and only when enough material has been assembled, it falls onto the black hole to give observable X rays. Hence, the X-ray source is transient. Recent observation of a large space velocity of Cygnus X-1 (Nelemans *et al.*, 1999) suggests that it has evolved similarly to the transient sources, with the difference that the companion to the black hole is an $\sim$18 $M_\odot$ O star. The latter pours enough matter onto the accretion disk so that Cyg X-1 shines continuously. We plan to describe the evolution of Cyg X-1 in a future paper (Brown *et al.*, 2000a).

Table 1 is an abbreviated list of data on transient sources. A more complete table is given in Brown *et al.* (1999b). Two of the steady X-ray sources, in the LMC, have been omitted, because we believe the LMC to be somewhat special because of its low metallicity; also masses, etc., of these two are not as well measured. Of the others, 6 are main-sequence K stars, one is main-sequence M, and the other two have masses greater than the Sun. The masses given are geometric means of the maximum and minimum masses given by the observers. The distance a between the black hole and the optical (visible) star is greater for the heavier stars than for the K- and M stars (except the more evolved one of them) as was expected in Sec. 7.1 for the spiraling in of star B. The table also gives the radius of the Roche Lobe and the specific orbital angular momentum of star B.

Five K stars have almost identical distance $a \sim 5$ $R_\odot$, and also Roche Lobe sizes, ~ 1.0 $R_\odot$. These Roche Lobes can be filled by K stars on the main sequence. The same is true for the M star. Together, K and M stars cover the mass range from 0.3 to 1 $M_\odot$. The two heavier stars have Roche Lobes of 3 and 5 $R_\odot$ which cannot possibly be filled by main-sequence stars of mass $\sim$2 $M_\odot$. We must therefore assume that these stars are subgiants, in the Herzsprung gap. These stars spend only about 1% of their life as subgiants, so we must expect that there are many "silent" binaries in which the 2 $M_\odot$ companion has not yet evolved off the main sequence and sits well within its Roche lobe, roughly 100 times more. The time as subgiants is even shorter for more massive stars; this explains their absence among the transient sources.

Therefore we expect a large number of "silent partners"; stars of more than 1 $M_\odot$, still on their main sequence, which are far from filling their Roche Lobe and therefore do not transfer mass to their black hole partners. In fact, we do not see any reason why the companion of the black hole could not have any mass, up to the ZAMS mass of the progenitor of the black hole; it must only evolve following the formation of the black hole. It then crosses the Herzsprung gap in such a short time, less than the thermal time scale, that star A cannot accept the mass from the companion, so that common envelope evolution must ensue. If we include these 'silent partners' in the birth rate, assuming a flat mass ratio distribution, we enhance the total birth rate of black-hole binaries by a factor 25 over the calculations by Brown *et al.* (1999b).

On the lower mass end of the companions, there is only one M star. This is explained in terms of the model of Sec. 7.1 by the fact that the stars of low mass will generally spiral into the He core of star A, and will coalesce with A, see below Eq. (23), so no relic is left. (Since the core is left spinning rapidly, these complete merger cases could also be suitable GRB progenitors.) As the outcome of the spiral-in depends also on other factors, such as the initial orbital separation and the primary mass, one may still have an occasional survival of an M star binary (note that the one M star companion is M0, very nearly in the K star range).

The appearance of the black hole transient X-ray binaries is much like our expectation of the relic of the binary which has made a hypernova: a black hole of substantial mass, and an ordinary star, possibly somewhat evolved, of smaller mass. We expect that star B would stop at a distance a_f from star A which is greater if the mass of B is greater (see Sec. 7.1). This is just what we see in the black-hole binaries: the more massive companion stars ($\sim$2 $M_\odot$) are further from the black hole than the K stars. We also note that the estimated birth rate of these binaries is high enough for them to be the progenitors of GRB, even if only in a modest fraction of them the conditions for GRB powering are achieved.

7.4. *Nova Scorpii 1994 (GRO J1655-40)*

Nova Sco 1994 is a black hole transient X-ray source. It consists of a black hole of $\sim$7 $M_\odot$ and a subgiant of about 2 $M_\odot$. Their separation is 17 $R_\odot$. Israelian *et al.* (1999) have analyzed the spectrum of the subgiant and have found that the α-particle nuclei O, Mg, Si and S have abundances 6 to 10 times the solar value. This indicates that the subgiant has been enriched by the ejecta from a supernova explosion; specifically, that some of the ejecta of the supernova which preceded the present Nova Sco (a long time ago) were intercepted by star B, the present subgiant. Israelian *et al.* (1999) estimate an age since accretion started from the assumption that enrichment has only affected the outer layers of the star. We here reconsider this: the time that passed since the explosion of the progenitor of the black hole is roughly the main-sequence lifetime of the present subgiant companion, which given its mass of $\sim$2 $M_\odot$ will be about 1 Gyr. This is so much longer than any plausible mixing time in the companion that the captured supernova ejecta must by now be uniformly mixed into the bulk of the companion. This rather increases the amount of ejecta that we require the companion to have captured. (Note that the accretion rate in this binary is rather less than expected from a subgiant donor, though the orbital period leaves no doubt that the donor is more extended than a main-sequence star (Regös *et al.*, 1998). It is conceivable that the high metal abundance has resulted in a highly non-standard evolution of this star, in which case one might have to reconsider its age).

The presence of large amounts of S is particularly significant. Nomoto *et al.* (2000) have calculated the composition of a hypernova from an 11 $M_\odot$. CO core, see Fig. 4. This shows substantial abundance of S in the ejecta. Ordinary supernovae

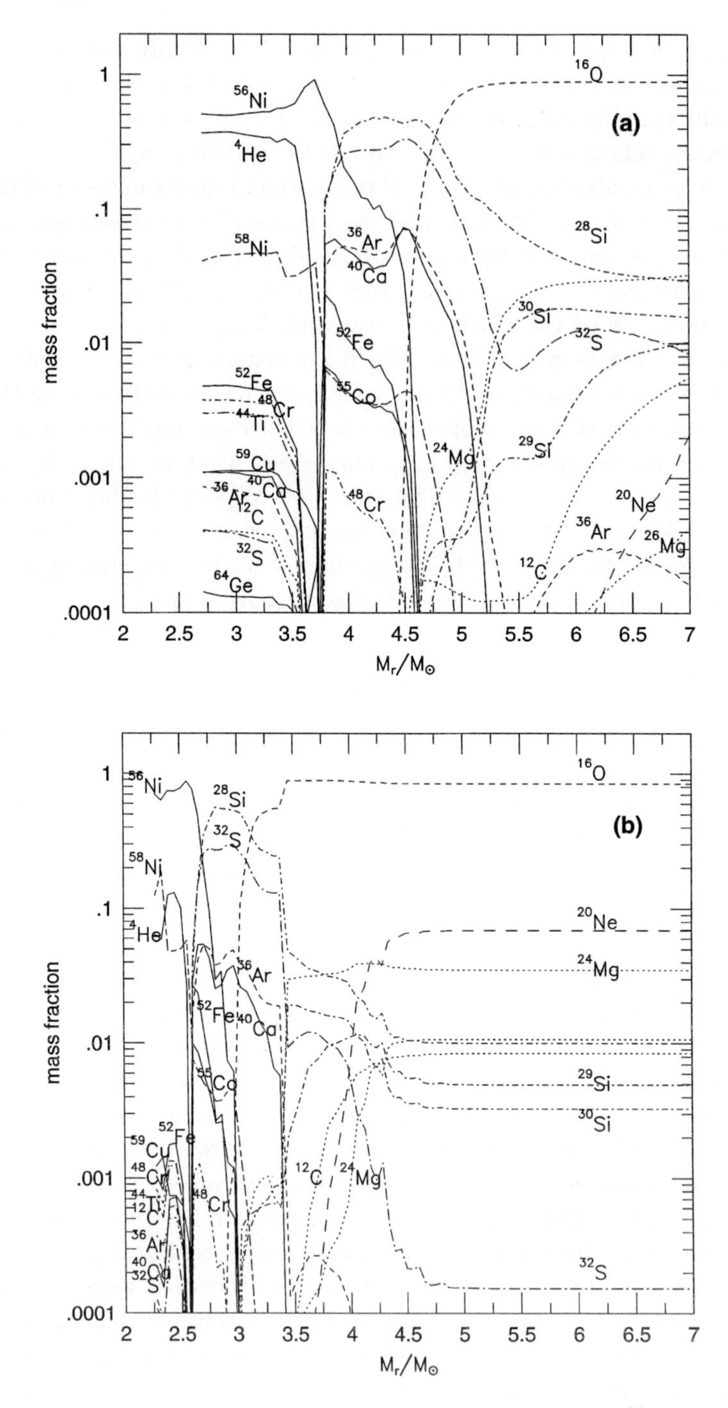

Fig. 4. The isotopic composition of ejecta of the hypernova ($E_K = 3 \times 10^{52}$ erg; upper) and the normal supernova ($E_K = 1 \times 10^{51}$ erg; lower) for a 16 $M_\odot$ He star, from Nomoto *et al.* (2000). Note the much higher sulphur abundance in the hypernova.

produce little of this element, as shown by the results of Nomoto *et al.* (2000) in Fig. 4. The large amount of S, as well as O, Mg and Si we consider the strongest argument for considering Nova Sco 1994 as a relic of a hypernova, and for our model, generally.

Figure 4 also shows that ^{56}Ni and ^{52}Fe are confined to the inner part of the hypernova, and if the cut between black hole and ejecta is about 5 $M_\odot$, there will be no Fe-type elements in the ejecta, as observed in Nova Scorpii 1994. By contrast hypernova 1998bw shows a large amount of Ni, indicating that in this case the cut was at a lower included mass.

The massive star A in Nova Sco will have gone through a hypernova explosion when the F-star B was still on the main sequence, its radius about 1.5 $R_\odot$. Since the explosion caused an expansion of the orbit, the orbital separation a was smaller at the time of the supernova than it is now, roughly by a factor

$$a_{\rm then} = a_{\rm now}/(1 + \Delta M/M_{\rm now})\,. \tag{32}$$

(ΔM is the mass lost in the explosion; see, e.g., Verbunt *et al.* (1990)). With $\Delta M \sim 0.8 M_{\rm now}$, as required by the high space velocity, this means $a_{\rm then} = 10\ R_\odot$. Therefore the fraction of solid angle subtended by the companion at the time of explosion was

$$\frac{\Omega}{4\pi} = \frac{\pi(1.5\ R_\odot)^2}{4\pi(10\ R_\odot)^2} \approx 6 \times 10^{-3}\,. \tag{33}$$

Assuming the ejecta of the hypernova to have been at least 5 $M_\odot$ (Nelemans *et al.*, 1999), the amount deposited on star B was

$$M_{\rm D} \gtrsim 0.03\ M_\odot\,. \tag{34}$$

The solar abundance of oxygen is about 0.01 by mass, so with the abundance in the F star being 10 times solar, and oxygen uniformly mixed, we expect $0.1 \times 2.5 = 0.25\ M_\odot$ of oxygen to have been deposited on the companion, much more than the total mass it could have captured from a spherically symmetry supernova. [Si/O] is 0.09 by mass in the Sun, and [Si/O] is 0.05, so since the over-abundances of all three elements are similar we expect those ratios to hold here, giving about 0.02 $M_\odot$ of captured Si and 0.01 $M_\odot$ of captured S. We therefore need a layer of stella ejecta to have been captured which has twice as much Si as S, at the same time as having about 10 times more than O. From Fig. 4, we see that this occurs nowhere in a normal supernova, but does happen in the hypernova model of Nomoto *et al.* (2000) at mass cuts of 6 $M_\odot$ or more. This agrees very nicely with the notion that a hypernova took place in this system, and that the inner 7 $M_\odot$ or so went into a black hole.

What remains is to explain how the companion acquired ten times more mass than the spherical supernova model allows, and once again we believe that the anser is given in recent hypernova calculations (MacFadyen and Woosley, 1999; Wheeler *et al.*, 2000): hypernovae are powered by jet flows, which means they are very asymmetric, with mass outflow along the poles being much faster and more energetic than along the equator. The disk provides a source for easily captured material in two ways: First, it concentrates mass in the equatorial plane, which will later be ejected mostly in that plane. Second, the velocity acquired by the ejecta is of the order of the propagation speed of the shock through it. This propagation speed is proportional to $\sqrt{P_2/\rho_1}$, where P_2 is the pressure behind the shock and ρ_1 the density ahead of it. The driving pressure will be similar in all directions (or larger, due to the jet injection, in the polar regions), whereas the disk density is much higher than the polar density. Hence, the equatorial ejecta will be considerably slower than even normal supernova ejecta, greatly increasing the possibility of their capture by the companion. Other significant effects of the disk/jet geometry are (1) that the companion is shielded from ablation of its outer layers by fast ejecta, which is thought to occur in spherical supernovae with companion stars (Marietta *et al.*, 2000) and (2) that there is no iron enrichment of the companion, because the iron — originating closest to the center — is either all captured by the black hole or ejected mainly in the jet, thus not getting near the companion (Wheeler *et al.* (2000); note that indeed no overabundance of Fe is seen in the companion of GRO J1655-40).

For the companion to capture the required 0.2–0.3 $M_\odot$ of ejecta it is sufficient that the ejecta be slow enough to become gravitationally bound to it. However, the material may not stay on: when the companion has so much mass added on a dynamical time scale it will be pushed out of thermal equilibrium, and respond by expanding, as do main-sequence stars that accrete mass more gradually on a time scale faster than their thermal time scale (e.g., Kippenhahn and Meyer-Hofmeister, 1977). During this expansion, which happens on a time scale much longer than the explosion, the star may expand beyond its Roche lobe and transfer some of its mass to the newly formed black hole. However, because the dense ejecta mix into the envelope on a time scale between dynamical and thermal, i.e., faster than the expansion time, this back transfer will not result in the bulk of the ejecta being fed back, though probably the material lose is still richer in heavy elements than the companion is now. Since the outer layers of the star are not very dense, and the mass transfer is not unstable because the black hole is much more massive than the companion, the total amount of mass transferred back is probably not dramatic. However, the expansion does imply that the pre-explosion mass of the companion was somewhat higher than its present mass, and that the amount of ejecta that needs to be captured in order to explain the abundances observed today is also somewhat higher than the present mass of heavy elements in the companion.

A further piece of evidence that may link Nova Sco 1994 to our GRB/hypernova scenario are the indications that the black hole in this binary is spinning rapidly. Zhang *et al.* (1997) argue from the strength of the ultra-soft X-ray component that the black hole is spining near the maximum rate for a Kerr black hole. However, studies by Sobczak *et al.* (1999) show that it must be spinning with less than 70% maximum. Gruzinov (1999) finds the inferred black hole spin to be about 60% of maximal from the 300 Hz QPO. Our estimates of the last section indicate that enough rotational energy will be left in the black hole so that it will still be rapidly spinning.

We have already mentioned the unusually high space velocity of -150 ± 19 km s^{-1}. Its origin was first discussed by Brandt *et al.* (1995), who concluded that significant mass must have been loss in the formation of the black hole in order to explain this high space velocity: it is not likely to acquire a substantial velocity in its own original frame of reference, partly because of the large mass of the black hole. But the mass lost in the supernova explosion is ejected from a moving object and thus carries net momentum. Therefore, momentum conservation demands that the center of mass of the binary acquire a velocity; this is the Blaauw–Boersma kick (Blaauw, 1961; Boersma, 1961). Note that the F-star companion mass is the largest among the black-hole transient sources, so the center of mass is furthest from the black hole and one would expect the greatest kick. Nelemans *et al.* (1999) estimate the mass loss in this kick to be 5–10 $M_{\odot}$.

In view of the above, we consider it well established that Nova So 1994 is the relic of a hypernova. We believe it highly likely that the other black-hole transient X-ray sources are also hypernova remnants. We believe it likely that the hypernova explosion was accompanied by a GRB if, as in GRB 980326, the energy was delivered in a few seconds. It is not clear what will happen if the magnetic fields are so low that the power is delivered only over a much longer time. There could then still be intense power input for a few seconds due to neutrino annihilation deposition near the black hole (Janka *et al.*, 1999), but that may not be enough for the jet to pierce through the He star and cause a proper GRB (MacFadyen and Woosley, 1999). At this point, we recall that the GRB associated with SN 1998bw was very sub-luminous, 10^5 times lower than most other GRB. While it has been suggested that this is due to us seeing the jet sideways, it is in our view more likely that the event was more or less spherical (Kulkarni *et al.*, 1998) and we see a truly lower-power event. A good candidate would be the original suggestion by Colgate (1968, 1974) of supernova shock break-out producing some gamma rays. Indications are that the expansion in SN 1998bw was mildly relativistic (Kulkarni *et al.*, 1998) or just sub-relativistic (Waxman and Loeb, 1999). In either case, what we may have witnessed is a natural intermediate event in our scenario: we posit that there is a continuum of events varying from normal supernovae, delivering 1 foe more or less spherically in ten seconds, to extreme hypernovae/GRB that deliver 100 foes in a highly directed beam. In the middle, there will be cases where the beam cannot pierce through the star, but the total energy delivered is well above a supernova, with as net result a hypernova accompanied by a very weak GRB.

7.5. *Numbers*

Nearly all observed black hole transient X-ray sources are within 5 kpc of the Sun. Extrapolating to the entire Galaxy, a total of 8,800 black-hole transients with main-sequence K companions has been suggested (Brown *et al.*, 1999b).

The lifetime of a K star in a black hole transient X-ray source is estimated to be $\sim 10^{10}$ yr (Van Paradijs, 1996) but we shall employ 10^9 yr for the average of the K-stars and the more massive stars, chiefly those in the "slient partners". In this case the birth rate of the observed transient sources would be

$$\lambda_K = 10^4/10^9 = 10^{-5} \text{ per galaxy yr}^{-1} . \tag{35}$$

We see no reason why low-mass companions should be preferred, so we assume that the formation rate of binaries should be independent of the ratio

$$q = M_{B,i}/M_{A,i} . \tag{36}$$

In other discussions of binaries, e.g., in (Portegies Zwart and Yungelson, 1998), it has often been assumed that the distribution is uniform in q. This is plausible but there is no proof. Since all primary masses M_A are in a narrow interval, 20 to 35 $M_\odot$, this means that M_B is uniformly distributed between zero and some average M_A, let us say 25 $M_\odot$. Then the total rate of creation of binaries of our type is

$$\lambda = \frac{25}{0.7}\lambda_K = 3 \times 10^{-4} \text{ galaxy}^{-1} \text{ yr}^{-1} . \tag{37}$$

This is close to the rate of mergers of low mass black holes with neutron stars which Bethe and Brown (1998) have estimated to be

$$\lambda_m \simeq 2 \times 10^{-4} \text{ galaxy}^{-1} \text{ yr}^{-1} . \tag{38}$$

These mergers have been associated speculatively with short GRBs, while formation of our binaries is supposed to lead to "long" GRBs (Fryer *et al.*, 1999). We conclude that the two types of GRB should be equally frequent, which is not inconsistent with observations. In absolute number both of our estimates Eqs. (37) and (38) are substantially larger than the observed rate of 10^{-7} galaxy^{-1} yr^{-1} (Wijers *et al.*, 1998); this is natural, since substantial beaming is expected in GRBs produced by the Blandford–Znajek mechanism. Although we feel our mechanism to be fairly general, it may be that the magnetic field required to deliver the BZ energy within a suitable time occurs in only a fraction of the He cores.

8. Discussion and Conclusion

Our work here has been based on the Blandford–Znajek mechanism of extracting rotational energies of black holes spun up by accreting matter from a helium star. We present it using the simple circuitry of "*The Membrane Paradigm*" (Thorne *et al.*, 1986). Energy delivered into the loading region up the rotational axis of the black hole is used to power a GRB. The energy delivered into the accretion disk powers a SN Ib explosion.

We also discussed black-hole transient sources, high-mass black holes with low-mass companions, as possible relics for both GRBs and Type Ib supernova explosions, since there are indications that they underwent mass loss in a supernova explosion. In Nova Sco 1994 there is evidence from the atmosphere of the companion star that a very powerful supernova explosion ('hypernova') occurred.

We estimate the progenitors of transient sources to be formed at a rate of 300 GEM (Galactic Events per Megayear). Since this is much greater than the observed rate of GRBs, there must be strong collimation and possible selection of high magnetic fields in order to explain the discrepancy.

We believe that there are strong reasons that a GRB must be associated with a black hole, at least those of duration several seconds or more discussed here. Firstly, neutrinos can deliver energy from a stellar collapse for at most a few seconds, and sufficient power for at most a second or two. Our quantitative estimates show that the rotating black hole can easily supply the energy as it is braked, provided the ambient magnetic field is sufficiently strong. The black hole also solves the baryon pollution problem: we need the ejecta that give rise to the GRB to be accelerated to a Lorentz factor of 100 or more, whereas the natural scale for any particle near a black hole is less than its mass. Consequently, we have a distillation problem of taking all the energy released and putting it into a small fraction of the total mass. The use of a Poynting flux from a black hole in a magnetic field (Blandford and Znajek, 1977) does not require the presence of much mass, and uses the rotation energy of the black hole, so it provides naturally clean power.

Of course, nature is extremely inventive, and we do not claim that all GRBs will fit into the framework outlined here. We would not expect to see all of the highly beamed jets following from the BZ mechanism head on, the jets may encounter some remaining hydrogen envelope in some cases, jets from lower magnetic fields than we have considered here may be much weaker and delivered over longer times, etc., so we speculate that a continuum of phenomena may exist between normal supernovae and extreme hypernovae/GRBs. This is why we call our effort "A Theory of Gamma Ray Bursts" and hope that it will be a preliminary attempt towards systematizing the main features of the energetic bursts.

Acknowledgements

We would like to thank Stan Woosley for much useful information. Several conservations with Roger Blandford made it possible for us to greatly improve our paper, as did valuable comments from Norbert Langer. This work is partially supported by the U.S. Department of Energy Grant No. DE-FG02-88ER40388. CHL is supported by Korea Research Foundation Grant (KRF-2002-070-C00027). HKL is supported also in part by KOSEF Grant No. 1999-2-112-003-5 and by the BK21 program of the Korean Ministry of Education.

Appendix A. Estimates of $\epsilon_\Omega = \Omega_{\rm disk}/\Omega_{\rm H}$

We collect here useful formulas needed to calculate $\epsilon_\Omega = \Omega_{\rm disk}/\Omega_{\rm H}$. First of all

$$\Omega_{\rm H} = \frac{\tilde{a}}{1+\sqrt{1-\tilde{a}^2}}\left(\frac{c^3}{2MG}\right)$$
$$= \frac{\sqrt{2}\tilde{a}}{1+\sqrt{1-\tilde{a}^2}}\left(\frac{r}{R_{\rm Sch}}\right)^{3/2}\Omega_{\rm K}\,, \tag{A.1}$$

$$\Omega_{\rm disk} = \omega\Omega_{\rm K}\left[1+\tilde{a}\frac{GM}{c^2}\sqrt{\frac{GM}{c^2r^3}}\right]^{-1}$$

$$= \omega\Omega_{\rm K}\left[1+\tilde{a}\left(\frac{R_{\rm sch}}{2r}\right)^{3/2}\right]^{-1}, \qquad (A.2)$$

where $\Omega_{\rm K} \equiv \sqrt{GM/R^3}$ and ω is dimensionless parameter ($0 < \omega < 1$). Thus

$$\frac{\Omega_{\rm disk}}{\Omega_{\rm H}} = \omega\frac{1+\sqrt{1-\tilde{a}^2}}{\sqrt{2}\tilde{a}} \times \left(\frac{R_{\rm Sch}}{r}\right)^{3/2}\left[1+\tilde{a}\left(\frac{R_{\rm Sch}}{2r}\right)^{3/2}\right]^{-1}. \qquad (A.3)$$

The numerical estimates are summarized in Table 2 for various ω and radii.

Table 2. Estimates of $\epsilon_\Omega = \Omega_{\rm disk}/\Omega_{\rm H}$ as a function of spin parameter and radius, where $r_{\rm mb}$ is the marginally bound radius and $r = r_{\rm ms}$ the marginally stable radius.

		$\Omega_{\rm disk}/\Omega_{\rm H}$			
ω	$\tilde{a}$	$r = r_{\rm mb}(\tilde{a})$	$r = r_{\rm ms}(\tilde{a})$	$r = 2R_{\rm Sch}$	$r = 3R_{\rm Sch}$
1.0	0.80	1.00	0.69	0.45	0.26
0.9	0.72	0.99	0.63	0.49	0.27
0.8	0.64	0.93	0.58	0.51	0.29
0.7	0.56	0.89	0.54	0.53	0.30
0.6	0.48	0.84	0.50	0.55	0.31
0.5	0.40	0.80	0.46	0.57	0.32

Appendix B. Spin-up of black holes by accretion

The specific angular momentum and energy of test particles in Keplerian circular motion, with rest mass δm, are

$$\tilde{E} \equiv \frac{E}{\delta m} = c^2\left[\frac{r^2 - R_{\rm Sch}r + a\sqrt{R_{\rm Sch}r/2}}{r(r^2-\frac{3}{2}R_{\rm Sch}r + a\sqrt{2R_{\rm Sch}r})^{1/2}}\right],$$

$$\tilde{l} \equiv \frac{l}{\delta m} = c\sqrt{\frac{R_{\rm Sch}r}{2}}\left[\frac{(r^2 - a\sqrt{2R_{\rm Sch}r}+a^2)}{r(r^2-\frac{3}{2}R_{\rm Sch}r + a\sqrt{2R_{\rm Sch}r})1/2}\right], \qquad (B.1)$$

where $R_{\rm Sch} = 2GM/c^2$ and BH spin $a = J/Mc = \tilde{a}(GM/c^2)$. The accretion of δm changes the BH's total mass and angular momentum by $\Delta M = \tilde{E}\delta m$ and $\Delta J = \tilde{\delta} m$.

Table 3. Properties of Schwarzschild and Kerr BH. (a) $r_{\rm lso} = r_{\rm ms}$ case: 6% (42%) of energy can be released during the spiral-in for Schwarzschild (maximally-rotating Kerr) BHs. (b) $r_{\rm lso} = r_{\rm mb}$ case: The released energy during the spiral-in is almost zero.

	a [GM/c^2]	$\tilde{l}$ [GM/c]	r [$R_{\rm Sch}$]	$\tilde{E}$ [c^2]
$r_{\rm ms}$	0	$2\sqrt{3} \approx 3.46$	3	$\sqrt{8/9} \approx 0.943$
	1	$2\sqrt{3} \approx 1.15$	1	$\sqrt{1/3} \approx 0.577$
$r_{\rm mb}$	0	4	2	1
	1	2	1	1

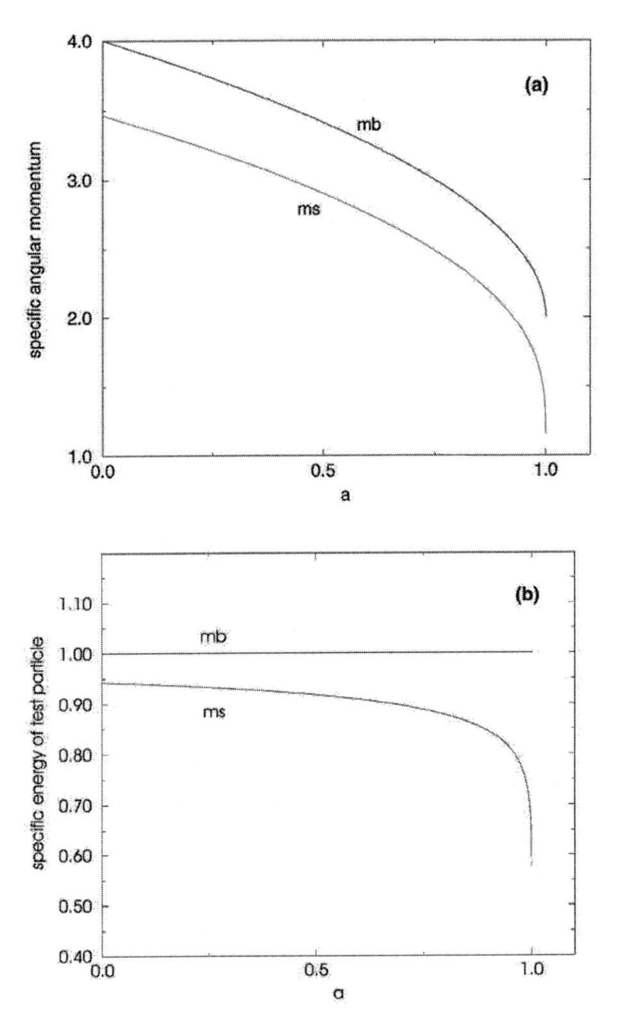

Fig. 5. Specific angular momentum and energy of test particle in units of [GM/c] and [c^2]. BH spin a is given in unit of [GM/c^2]. For the limiting values at $a = 0$ and GM/c^2, refer Table 3.

The radii of marginally bound ($r_{\rm mb}$) and stable ($r_{\rm ms}$) orbits are given as

$$r_{\rm mb} = R_{\rm Sch} - a + \sqrt{R_{\rm Sch}(R_{\rm Sch} - 2a)}\,,$$

$$r_{\rm ms} = \frac{R_{\rm Sch}}{2}\{3 + Z_2 - [(3 - Z_1)(3 + Z_1 + 2Z_2)]^{1/2}\}\,,$$

$$Z_1 = 1 + \left(1 - \frac{4a^2}{R_{\rm Sch}^2}\right)^{1/3} \times \left[\left(1 + \frac{2a}{R_{\rm Sch}}\right)^{1/3} + \left(1 - \frac{2a}{R_{\rm Sch}}\right)^{1/3}\right]\,,$$

$$Z_2 = \left(3\frac{4a^2}{R_{\rm Sch}^2} + Z_1^2\right)^{1/2}\,. \tag{B.2}$$

The numerical values of the specific angular momentum and energy of test particles are summarized in Table 3 and Fig. 5. In Fig. 6, we test how much mass we need in order to spin up the non-rotating black hole up to given $\tilde{a}$. Note that the last stable orbit is almost Keplerian even with the acretion disk, and we assume 100% efficiency of angular momentum transfer from the last stable Keplerian orbit to BH. In order to spin-up the BH up to $\tilde{a} = 0.9$, we need ~68% (52%) of original non-rotating BH mass in case of $r_{\rm lso} = r_{\rm ms}$ ($r_{\rm mb}$). For a very rapidly rotating BH with $\tilde{a} = 0.99$, we need 122% and 82%, respectively. For $r_{\rm lso} = r_{\rm ms}$, there is an upper limit, $\tilde{a} = 0.998$, which can be obtained by accretion (Thorne, 1974). In the limit where $r_{\rm lso} = r_{\rm mb}$, however, spin-up beyond this limit is possible because the photons can be captured inside thick accretion disk, finally into BH (Abramowicz *et al.*, 1988).

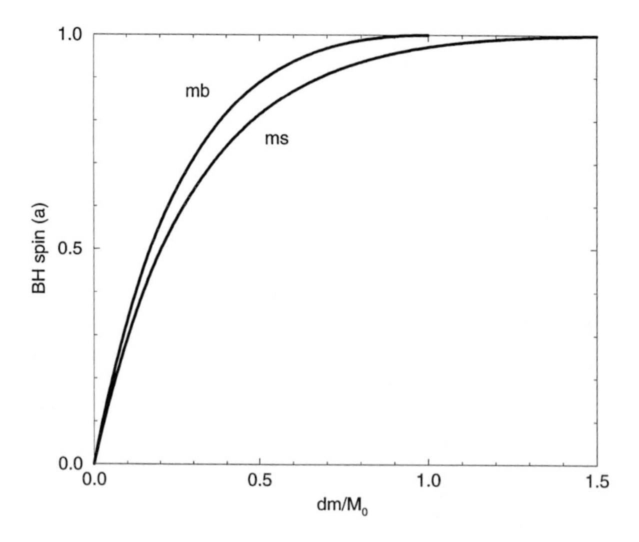

Fig. 6. Spinning up of Black Holes. The BH spin a is given in units of [GM/c^2] and δm is the total rest mass of the accreted material.

References

Abramowicz M. A., Czerny B., Lasota J. P. and Szuszkiewicz E., *ApJ* **332**, 646 (1988).

Andersen M. I., Castro-Tirado A. J., Hjorth J., Moller P., Pedersen H., Caon N., Marina Cairos L., Korhonen H., Zapatero Osorio M. R., Perez E. and Frontera F., *Sci.* **283**, 2075 (1999).

Bethe H. A., *Rev. Mod. Phys.* **62**, 801 (1990).

Bethe H. A. and Brown G. E., *ApJ* **506**, 780 (1998).

Bhattacharya D., *JApA* **11**, 125 (1990).

Bhattacharya D., in: J. Ventura and D. Pines (Eds.), *Neutron Stars: Theory and Observation*, NATO ASI Ser. C., vol. 344 (1991), p. 103.

Blaauw A., *Bull. Astron. Inst. Netherlands* **15**, 265 (1961).

Blandford R. D. and Spruit H., (2000), in preparation.

Blandford R. D. and Znajek R. L., *MNRAS* **179**, 433 (1977).

Bloom J. S. *et al.*, *Natur.* **401**, 453 (1999).

Boersma J., *Bull. Astron. Inst. Netherlands* **15**, 291 (1961).

Brandenburg A., Nordlund Å., Stein R. F. and Torkelsson Y., *ApJ* **458**, L45 (1996).

Brandt W. N., Podsiadlowski P. and Sigurdsson S., *MNRAS* **277**, L35 (1995).

Brown G. E. and Bethe H. A., *ApJ* **423**, 659 (1994).

Brown G. E., Weingartner J. C. and Wijers R. A. M. J., *ApJ* **463**, 297 (1996).

Brown G. E., Lee C.-H., Wijers R. A. M. J. and Bethe H. A., *Phys. Rep.* (1999a), in press, astro-ph/9910088.

Brown G. E., Lee C.-H. and Bethe H. A., *NewA* **4**, 313 (1999b).

Brown G. E., Heger A., Langer N., Lee C.-H., Wellstein S. and Bethe H. A., (2000a), in preparation.

Brown G. E., Lee C.-H. and Bethe H. A., *ApJ*, to be published, (2000b), astro-ph/9909132.

Colgate S. A., *Can. J. Phys.* **46**, 5471 (1968).

Colgate S. A., *ApJ* **187**, 333 (1974).

Eichler D., Livio M., Piran T. and Schramm D., *Natur.* **340**, 126 (1989).

Fryer C. L. and Woosley S. E., *ApJ* **502**, L9 (1998).

Fryer C. L., Woosley S. E. and Hartmann D. H., *ApJ* **526**, 152 (1999).

Galama T. *et al.*, *Natur.* **395**, 670 (1998).

Galama, T. *et al.*, *ApJ* **536**, 185 (2000).

Gruzinov A., *ApJ* **517**, L105 (1999).

Heger A. and Wellstein S., in website of A. Heger (2000).

Heger A., Langer N. and Woosley S., *ApJ* **528**, 368 (2000).

Höflich P., Wheeler J. C. and Wang L., *ApJ* **521**, 179 (1999).

Israelian G., Rebolo R., Basri G., Casares J. and Martin E. L., *Natur.* **401**, 142 (1999).

Iwamoto K. *et al.*, *Natur.* **395**, 672 (1998).

Janka H. T., Ebert T., Ruffert M. and Fryer C. L., *ApJ* **527**, L39 (1999).

Kippenhahn R. and Meyer-Hofmeister E., *A&A* **54**, 539 (1977).

Kouveliotou C. *et al.*, *Natur.* **393**, 235 (1998).

Kouveliotou C. *et al.*, *ApJ* **510**, L115 (1999).

Kulkarni S. R., Frail D. A., Wieringa M. H., Ekers R. D., Sadler E. M., Wark R. M., Higdon J. L., Phinney E. S. and Bloom J. S., *Natur.* **395**, 663 (1998).

Lattimer J. M. and Prakash M., *ApJ* (2000), submitted, astro-ph/0002232.

Lee H.-K., Wijers R. A. M. J. and Brown G. E., *Phys. Rep.* **325**, 83 (1999).

Lee H.-K., Brown G. E. and Wijers R. A. M. J., *ApJ* **536**, 416 (2000).

Li L.-X., *PhRvD* **61**, 084016 (2000).

Lindblom L. and Owen B. J., *PhRvL* **80**, 4843 (1999).

Livio M. and Pringle J. E., *ApJ* **505**, 339 (1998).

Lovelace R. V. E., MacAuslan J. and Burns M., in: *AIP Conf. Ser. 56: Particle Acceleration Mechanisms in Astrophysics*, (1979), p. 399.
MacDonald D. and Thorne K. S., *MNRAS* **198**, 345 (1982).
MacFadyen A. I. and Woosley S. E., *ApJ* **524**, 262 (1999).
Marietta E., Burrow A. and Fryxell B., *ApJ* (2000), in press, astro-ph/9908116.
Narayan R. and Yi I., *ApJ* **428**, L13 (1994).
Nelemans G., Tauris T. M. and van den Heuvel, E. P. J., *A&A* **352**, L87 (1999).
Nomoto K., Mazzali P. A., Nakamura T., Iwamoto K., Maeda K., Suzuki T., Turatto M., Danziger I. J. and Patat F., in: M. Livio, N. Panagia and K. Sahu (Eds.), *The Greatest Explosions since the Big Bang: Supernovae and Gamma-Ray Bursts*, CUP: Cambridge, (2000), astro-ph/0003077.
Phinney E. S., Ph.D. Thesis, University of Camridge, Cambridge, UK (1983).
Portegis Zwart S. F. and Yungelson L. R., *A&A* **332**, 173 (1998).
Rasio F. A. and Livio M., *ApJ* **471**, 366 (1996).
Regõs E., Tout C. A. and Wickramasinghe D., *ApJ* **509**, 362 (1998).
Shakura N. I. and Sunyaev R. A., *A&A* **24**, 337 (1973).
Sobczak G. J., McClintock J. E., Remillard R. A., Bailyn C. D. and Orosz J. A., *ApJ* **520**, 776 (1999).
Spruit H. and Phinney E. S., *Natur.* **393**, 139 (1998).
Thorne, K. S., *ApJ* **191**, 507 (1974).
Thorne K. S., Price R. H. and MacDonald D. A., *The Membrane Paradigm*, Yale University Press, New Haven (1986).
Torkelsson U., Brandenburg A., Nordlund A. and Stein R. F., *Astron. Lett. Comm.* **34**, 383 (1996).
Van Paradijs J., *ApJ* **494**, L45 (1996).
Verbunt F. W. M., Wijers R. A. M. J. and Burn H. M. G., *A&A* **234**, 195 (1990).
Waxman E and Loeb A., *ApJ* **515**, 721 (1999).
Wellstein S. and Langer N., *A&A* **350**, 148 (1999).
Wheeler J. C., Höflich P., Wang L. and Yi I., in: *Proc. Fifth Huntsville Conf. Gamma-Ray Bursts*, (2000), astro-ph/9912080.
Wijers R. A. M. J., Bloom J. S., Bagla J. S. and Natarajan P., *MNRAS* **294**, L13 (1998).
Woosley S. E., *ApJ* **350**, 218 (1988).
Woosley S. E., *ApJ* **405**, 273 (1993).
Woosley S. E. and Weaver T. A., *ApJS* **101**, 181 (1995).
Zhang S. N., Cui W. and Chen W., *ApJ* **482**, L155 (1997).

M. Veltman

RECONSIDERATION OF A QUESTION POSED BY FRANK YANG AT THE 1973 BONN CONFERENCE

M. VELTMAN
Emeritus MacArthur Professor
University of Michigan, USA

Bonn 1973. Questions after the talk by M. Veltman

Yang: The recent activities are very exciting in this field. By combining the concept of gauge fields and the concept of broken symmetry, it was proved — and you say it is a closed chapter — that there is renormalizability.

I would like to ask you, in your opinion, when the dust settles down, what essential parts of the recent developments are likely to remain, and what new ingredients, in a general way, are still needed for a succesful theory?

Veltman: There are several things that seem to be of deeper significance. First, electromagnetism and gravitation are gauge theories, and now there opens the real possibility that other interactions have also some intricate gauge symmetry. It appears that the gauge idea, the Yang–Mills structure, is a fundamental property. But note that these theories owe their interest to the fact that they are renormalizable. I do not know how fundamental this notion of renormalizability is, but it is really miraculous that these theories are renormalizable. This, of course, barring gravitation of which we do know very little. It seems to me that this can be no accident.

One of the things that appeals less to me is for instance the question of anomalies. It is strange to think that such far fetched technicalities would determine the structure of the world. There must be something that we have not understood about parity. Again and again it strikes me as something that fits not naturally in the scheme.

Another such thing is the Higgs mechanism. Despite its real beauty in various circumstances it hurts that one must artificially introduce new particles. Perhaps there are ways out, and one can imagine that the Higgs system is a way to describe an effective lagrangian at low energies. In the literature there are already attempts in that direction, viz. Coleman and E. Weinberg.

Perhaps a deeper understanding of strong interactions is necessary before we can make further progress. Perhaps also more can be learned by studying gravitation, which after all exists too. Who knows!

When Frank Yang poses a question it is rarely something that you can quickly do away with. Indeed, in some real sense I have been busy with the answer almost to this day.

The questions as you see them here are from the written form that we both produced immediately after the question period. They are slightly different from the spoken versions, which I listened to as I have a recording of that talk and the question period. The differences are not significant, but I should perhaps mention them.

First, Frank opens his question with the statement:

"Knowing that the speaker oftentimes makes very frank remarks ..."

This of course is an invitation to say something gruesome, such as

"We now can forget about all that Regge pole nonsense ..."

and indeed, this was on my lips, but I did not say it. What I did say is that I believed that the frantic creation of models would stop; people may remember that at that time an immense number of papers on that subject were produced. Also, I did not mention the anomaly in my spoken answer. Other than that there is really not much difference between what we said and what we wrote. After all, we wrote it directly after the talk.

I should perhaps put the above in the context of the state of affairs at that moment. Quantum chromodynamics was waiting in the wings, but not there yet (as you can see from my response). Concerning gravitation, 't Hooft and I had just finished a paper on divergencies in gravitation. I had not yet made the cosmological constant connection to the Higgs system; that came to me a few months later, during a seminar by Deser in Orsay. Else I would most certainly have mentioned that the vanishingly small cosmological constant as observed is hard to understand in view of the Higgs creating a huge contribution to that constant. And indeed, that fact has very much prejudiced me against the Higgs system as such. Something has come out of that:

- The speculation that the real world can perhaps approximately be described by a Higgs system with a very large Higgs mass;
- The requirement that quadratic divergencies (as found in the cosmological constant) cancel (on the one-loop level).

That cancellation would also happen in case of a super-symmetry. This requirement produces a sum rule involving the masses of the quarks, leptons and vector bosons; given that the top quark and the Higgs are very heavy they dominate in that equation and the prediction is that the Higgs particle has a mass about twice that of the top mass. That may be true. So far we do not know. Present data suggest a lower Higgs mass, but with large error bars. Note that so far undiscovered particles would also enter in the sum-rule, so we are not dealing here with a definite equation. If however the equation holds as stated, involving only the presently known particles, then we have a truly interesting situation.

In any case, I still stand by my remarks on the Higgs system. Added to that is the consequence for gravitation, i.e. the cosmological constant. Something is very wrong there!

But let us go back. Did the dust settle? Some of it indeed settled, but I certainly would have been wrong if I had stated that the Regge pole nonsense would disappear. One might say that the Regge pole stuff evolved into the present day string theories. Now let us get one thing straight: so far no physics has come out of string theory. So indeed, it may all be nonsense. But the dust has not settled there. I would however be tempted to say that in my opinion, in due time, string theories will disappear. My reason for this is that these theories are evolving without any impetus from physical reality. There is nothing, but nothing, that points in that direction. This in contrast for example with gauge theories around 1970: the validity of current equations, CVC, PCAC, current algebra, the Adler–Weisberger relation, clearly pointed to a gauge theory.

Also the research into gravitation has not produced any physics result. The extensive discussions about black holes sound to me like arguments about the number of angels that can dance on the point of a pin. So here, Frank, you have some frank opinions! But to be fair: there has been very little progress altogether. It is not that easy to guess what nature is doing, and without new experimental data there will very likely be no progress. To quote Lederman on theorists: they need help!

That brings us now to the question of renormalizability. Very little progress has been made on that issue even if it has been with us for more than half a century. How fundamental is it? Is it a guiding principle? I would like to enter into a discussion on that.

Most theorists would probably agree that if a theory is renormalizable, it is still not acceptable. Even if the infinities are 'invisible' they really should not be there. Of course, today the infinities of the self-energies of the elementary particles are responsible for the fact that they must be taken as free parameters (in which we can hide the infinity), and consequently we cannot calculate those masses. The same holds for coupling constants. The large number of parameters in the Standard Model is truly embarrassing, in addition to the fact that we have absolutely no inkling as to why there are three identical fermion generations and so on. Only the Z_0 mass is determined if we assume that the Higgs system is the simplest possible ($\rho = 1$). Beyond that only $SU5$ has produced some statements on some of the free parameters. That is why I am still sympathetic to at least the phenomenology of $SU5$. It also suggests massless neutrino's, which is were I put my bets these days.

There is one interesting remark that ties in with the famous question of parity violation. Why is parity violated? To that question some answer can be given.

Basically, with the advent of the Higgs mechanism we have two independent ways of generating, theoretically, particle masses. First, we can just introduce a mass term in the Lagrangian, and secondly, additional mass may be generated by the Higgs mechanism. Here now is this simple observation:

- If the particle masses are exclusively due to the Higgs mechanism then parity must be violated.

The reason is that the Higgs field is not an $SU2 \times U1$ singlet, it is a doublet. If now an invariant coupling must be constructed involving the Higgs field and the left- and right-handed fermion fields (that is the type of coupling needed to generate a mass term if the Higgs field obtains a vacuum expectation value) then this is not possible if the left- and right-handed fermion fields are of the same multiplet type with respect to the $SU2 \times U1$. In other words, left- and right-handed fermions are coupled differently to the vector bosons. Presto, parity is violated.

As it happens, the masses of all particles in the Standard Model are produced by the Higgs mechanism. There is no mass put by hand. I consider this an important clue: whatever the Higgs system, it appears to be the Deus ex machina of the masses of all particles. In some vague sense that makes us think again of gravitational interactions.

Let me now go back to the question of renormalizabilty. The idea that there should be no infinities altogether is still sympathetic to me. This raises very interesting questions. If the infinities must disappear, then somehow nature uses a cut-off mechanism concerning the infinities. In the pre-gauge theory days many people thought that the infinities of quantum-electrodynamics would disappear because the electron was not point-like but had a small size. Note that this view fits in with the idea that all physics is embedded in four-dimensional space-time.

A finite size is however not tenable in gauge theories such as the Standard Model. In a gauge theory there are many Ward identities, relating for example the coupling of an electron to the vector bosons to the coupling of the vector bosons among themselves and vice versa. It is, or so I think, not possible to introduce a finite electron size compatible with gauge invariance. So that possibility disappears. What now?

It is hard to say how nature could handle this. String people claim that some of their theories would make everything finite, but again, no physics has been produced by string theories. Let us try another approach.

Computing radiative corrections in a gauge theory necessitates the introduction of a mathematical cut-off scheme. In the old days of quantum-electrodynamics that was the Pauli–Villars scheme. That scheme is not compatible with non-abelian gauge invariance and another scheme had to be invented.

Now you might argue that on the one hand nature must use some cut-off scheme, while on the other hand we may have one or more mathematical cut-off schemes. It is tempting to say that perhaps one of our current schemes is close to physical reality. With the old Pauli–Villars scheme that was hard to do, because it involved particles with an unphysical metric. What is available today?

Well, essentially there is only one gauge invariant scheme, which is dimensional regularization. Can we imagine some physical realization of that scheme? In other words, let us take that scheme very seriously. Even if it does not correspond precisely to physical reality we may learn from the exercise. If you want some analog: the Higgs system can be considered as a cut-off scheme for a massive Yang–Mills theory. This particular cut-off scheme can be interpreted as real physics.

Here we get into real trouble. First of all parity violation and the anomaly are awkward in that scheme. The ubiquitous γ^5 cannot be generalized to continuous dimensions. In calculations beyond one loop that is really painful. Also, external fermion lines involve spinors that cannot be generalized to continuous dimensions either. In fact, concerning fermions, the only thing that can be generalized are traces. Here I have two observations.

First, concerning γ^5, I have satisfied myself (although I have no strict proof) that the following rule holds. If there are no anomalies then γ^5 can formally be assumed to anti-commute with all other γ matrices. In other words, that is how you handle γ^5 when computing traces. That removes a lot of trouble in doing calculations. Nonetheless, the presence of γ^5 is a constant reminder that we are in strictly four dimensions, and I see this as a really difficult point in this context.

Secondly, concerning spinors, one can reformulate the current theory such that it involves only traces. I will not go into detail; the essential point is that after squaring the amplitudes everything reduces to traces. In the end you do not have to go that far, but that is the basis. However, it turns out that this introduces ambiguities in case of through-going fermion lines. Further consideration shows, luckily, that these ambiguities do not appear in a renormalizable theory. But you cannot use dimensional regularization to make non-renormalizable theories involving fermions finite! It is interesting to note that this subject has been of relevance with respect to theoretical work on 'heavy quarks'. In such work one assumes certain quarks to be very heavy, and certain parts of diagrams reduce to the well-known four-fermion couplings. There has been considerable argument on how to work with these theories, and that is really due to the ambiguities mentioned. I have not followed that work, and I do not know if this problem has been settled, which should be possible since one starts from a renormalizable theory.

There is however an even more disconcerting point about dimensional regularization. It cannot be formulated in ordinary space-time, and also not on the Lagrangian level. This is different for the case of the Pauli–Villars scheme: one can write a Lagrangian that produces the desired result (involving some negative metric fields). But dimensional regularization can only be formulated for diagrams, in momentum space. You cannot formulate it in coordinate space. The scheme has nothing to do with going to continuous dimensions in space-time. Space-time can be seen as the Fourier transform of momentum space, but a continuous dimension in momentum space does not translate to a continuous dimension in space-time.

Well, you might say, so what. As long as any physical result can be Fourier transformed to coordinate space that is an irrelevant observation. But consider: while unitarity (conservation of probability) is a simple property in momentum space, causality is not. Perhaps all problems that we have with causality in quantum mechanics is due to our insistence that particles literally move in ordinary space and time.

As long as we are dealing with perturbation theory there is no difficulty with Fourier transforming from momentum space to coordinate space. The problem starts if we are dealing with singular configurations and boundary conditions. If we take the view that physics is really based in momentum space then it is not reasonable to impose coordinate space boundary conditions. And if some configuration cannot be Fourier transformed because of its singular behavior, then it is unclear how to interpret the results. A case in point are black holes. You cannot Fourier transform them. It is entirely unclear how these objects look like in momentum space. So, perhaps black holes as found in the classical theory of gravitation do not exist. In actual fact, when one considers the astrophysical claims concerning the existence of black holes you find that there is no real hard evidence. To me, hard evidence would be to show that a certain amount of mass is concentrated in an area smaller than the associated Schwarzschild radius. All papers on that subject have usually a sentence of the form "the only plausible explanation is a single black hole". But in physics things do not work as in a Sherlock Holmes story: eliminate all other suspects and you have the culprit. There is a lot of dust in that area!

Concerning boundary conditions, if we think that momentum space is the basic entity then that solves in first instance the cosmological constant problem. It is a boundary value problem. That is the inverse of the customary view: no one worries about a cosmological constant in momentum space, and no one would dream of imposing boundary conditions in momentum space. However, the problem comes back in another form in momentum space, and so far that has not been solved.

All in all, the dust has not settled. It probably never will!

L. Alvarez-Gaume

DIRECT $\mathcal{T}$-VIOLATION MEASUREMENTS AND $\mathcal{T}$-ODD EFFECTS IN DECAY EXPERIMENTS*

L. ALVAREZ-GAUMÉ[†], C. KOUNNAS[†,‡], S. LOLA[†] and P. PAVLOPOULOS[§]

[†]*CERN Theory Division, CH-1211 Geneva, Switzerland*
[‡]*Ecole Normale Supérieure 24 rue Lhommond, F-75231, Paris Cedex 05, France*
[§]*Institut für Physik, University of Basle CH-4056, and*
CPLEAR Collaboration, CH-1211 Geneva, Switzerland

Motivated by the recent experimental announcements for direct measurements of time-reversal non-invariance in the neutral kaon system, we make a comparative discussion of the CPLEAR and KTeV measurements. The most suitable way to consistently incorporate the mixing, the time evolution and the decays of kaons, is to describe the neutral kaon system as a system with a non-Hermitean Hamiltonian. In this framework, the physical (decaying) incoming and outgoing states are distinct and belong to dual spaces. Moreover, since they are eigenstates of the full Hamiltonian, they never oscillate. This is directly manifest in the orthogonality conditions of the physical states, which entirely determine the evolution of the kaon system. Along these lines we conclude: CPLEAR studies K^0–$\bar{K}^0$ oscillations, a process where initial and final states can be reversed, the CPLEAR asymmetry being an effect directly related to the definition of time-reversal. Conclusively, CPLEAR provides a direct measurement of $\mathcal{T}$-violation without any assumption either on unitarity or on $\mathcal{CPT}$-invariance. The KTeV experiment studies in particular the process $K_L \to \pi^+\pi^-e^+e^-$, where they measure a $\mathcal{T}$-odd effect. However, using unitarity together with estimates of the final state interactions, it should be possible to determine whether this effect can be identified with a genuine $\mathcal{T}$-reversal violation.

1. Introduction

Recently, the CPLEAR experiment at CERN, reported the first direct observation of time-reversal violation in the neutral kaon system.[1] This observation is made by comparing the probabilities of a $\bar{K}^0$ state transforming into a K^0 and vice-versa. Moreover, the KTeV experiment at Fermilab, similarly reported evidence for $\mathcal{T}$-violation in the decay $K_L \to \pi^+\pi^-e^+e^-$.[2] In the present note, we will discuss the experimental asymmetries used by both collaborations and interpret their measurements on $\mathcal{CP}$, $\mathcal{T}$ and/or $\mathcal{CPT}$-violation.

The discrete symmetry properties of the neutral kaon system have been extensively studied in the literature.[4] To analyze this issue, in a consistent way one needs to study a system with a non-Hermitean Hamiltonian. This is clear, because of the

*Talk given by S. Lola at the XXXIVth Rencontres de Moriond on Electroweak Interactions and Unified Theories, Les Arcs, 13–20 March 1999.

following: Although the physical kaons at rest coincide with the strong interaction (strangeness) eigenstates $|K^0\rangle = |d\bar{s}\rangle$ and $|\bar{K}^0\rangle = |\bar{d}s\rangle$, the latter are not the eigenstates of the full Hamiltonian. Since however weak interactions do not conserve strangeness (but also allow K^0–$\bar{K}^0$ oscillations) the full Hamiltonian eigenstates, denoted by $|K_S\rangle$ and $|K_L\rangle$, are different from the strangeness eigenstates, and obey the relations

$$\begin{aligned} H|K_S\rangle &= \lambda_S|K_S\rangle, \qquad & |K_S(t)\rangle &= e^{-i\lambda_S t}|K_S\rangle\,, \\ H|K_L\rangle &= \lambda_L|K_L\rangle, \qquad & |K_L(t)\rangle &= e^{-i\lambda_L t}|K_L\rangle\,, \end{aligned} \tag{1}$$

with $\lambda_L = m_L - i\Gamma_L/2$ and $\lambda_S = m_S - i\Gamma_S/2$, where $m_{S,L}$ denotes the masses of the physical kaons and $\Gamma_{S,L}$ their decay widths. The complexity of the eigenvalues, implies the non-Hermiticity of the full Hamiltonian of the neutral kaon system.

Non-Hermiticity of H implies that the physical incoming and outgoing states ($|K^{\rm in}_{S,L}\rangle$ and $|K^{\rm out}_{S,L}\rangle \equiv \langle K^{\rm out}_{S,L}|^\dagger$ respectively), are not identical, but instead belong to two distinct (dual) spaces.[3] In the Heisenberg representation (where the states are time-independent), the physical incoming and outgoing states coincide with the left- and right-eigenstates of the full Hamiltonian:

$$\begin{aligned} H|K^{\rm in}_{S,L}\rangle &= \lambda_{S,L}|K^{\rm in}_{S,L}\rangle, \qquad & \langle K^{\rm in}_{S,L}|H^\dagger &= \langle K^{\rm in}_{S,L}|\lambda^*_{S,L}\,, \\ H^\dagger|K^{\rm out}_{S,L}\rangle &= \lambda^*_{S,L}|K^{\rm out}_{S,L}\rangle, \qquad & \langle K^{\rm out}_{S,L}|H &= \langle K^{\rm out}_{S,L}|\lambda_{S,L}\,, \end{aligned} \tag{2}$$

where

$$|K^{\rm out}_{S,L}\rangle \equiv \langle K^{\rm out}_{S,L}|^\dagger \neq |K^{\rm in}_{S,L}\rangle, \qquad \langle K^{\rm in}_{S,L}| \equiv |K^{\rm in}_{S,L}\rangle^\dagger \neq \langle K^{\rm out}_{S,L}|\,. \tag{3}$$

Notice that *only if* $H = H^\dagger$, $\lambda_{S,L} = \lambda^*_{S,L}$ and $|K^{\rm out}_{S,L}\rangle = |K^{\rm in}_{S,L}\rangle$, thus the incoming and outgoing states are identical. In the generic case ($H \neq H^\dagger$), the time evolution of the incoming and outgoing states $|\Psi^{\rm in}_I(t_i)\rangle$ and $|\Psi^{\rm out}_I(t_f)\rangle$ are obtained from $|\Psi^{\rm in}_I\rangle$ and $|\Psi^{\rm out}_I\rangle$, using the evolution operators e^{-iHt_i} and $e^{-iH^\dagger t_f}$ respectively:

$$|K^{\rm in}_{S,L}(t_i)\rangle = e^{-iH\,t_i}|K^{\rm in}_{S,L}\rangle, \qquad |K^{\rm out}_{S,L}(t_f)\rangle = e^{-iH^\dagger\,t_f}|K^{\rm out}_{S,L}\rangle\,. \tag{4}$$

From the above equations, follows the evolution of the conjugate states:

$$\langle K^{\rm in}_{S,L}(t_i)| = \langle K^{\rm in}_{S,L}|e^{iH^\dagger\,t_i}, \qquad \langle K^{\rm out}_{S,L}(t_f)| = \langle K^{\rm out}_{S,L}|e^{iH\,t_f}\,. \tag{5}$$

An important point to stress here, is that the physical incoming and outgoing eigenstates have to obey *at all times* the orthogonality conditions[3]

$$\langle K^{\rm out}_I(t_f)|K^{\rm in}_J(t_i)\rangle = \langle K^{\rm out}_I|e^{-iH\Delta t}|K^{\rm in}_J\rangle = e^{-i\lambda_I\Delta t}\delta_{IJ}\,, \tag{6}$$

and in particular for $\Delta t = 0$

$$\begin{aligned} \langle K^{\rm out}_L|K^{\rm in}_S\rangle &= 0, \qquad & \langle K^{\rm out}_S|K^{\rm in}_L\rangle &= 0\,, \\ \langle K^{\rm out}_S|K^{\rm in}_S\rangle &= 1, \qquad & \langle K^{\rm out}_L|K^{\rm in}_L\rangle &= 1\,. \end{aligned} \tag{7}$$

unlike what has been stated in a wide part of the literature. These conditions express the fact that the Hamiltonian eigenstates cannot oscillate to each other at any time, and therefore an initial $|K^{\rm in}_S\rangle$ may not be transformed to a final $|K^{\rm out}_S\rangle$.

Moreover, it follows that the inner products among incoming (outgoing) states *do not obey* the usual orthogonality conditions

$$\langle K_I^{\rm in}|K_J^{\rm in}\rangle \neq \delta_{IJ} \quad \text{and} \quad \langle K_I^{\rm out}|K_J^{\rm out}\rangle \neq \delta_{IJ} \,. \tag{8}$$

Finally, in the basis of the states K_L and K_S, H can be expressed in terms of a diagonal 2×2 matrix

$$H = |K_S^{\rm in}\rangle\lambda_S\langle K_S^{\rm out}| + |K_L^{\rm in}\rangle\lambda_L\langle K_L^{\rm out}|\,, \tag{9}$$

where the unity operator **1** takes the form:

$$\mathbf{1} = \sum_{I=S,L} |K_I^{\rm in}\rangle\langle K_I^{\rm out}|\,. \tag{10}$$

2. Study of Discrete Symmetries in the Neutral Kaon System

Having clarified our formalism, we may now proceed to study particle–antiparticle mixing in the neutral kaon system. A convenient representation to study the action of $\mathcal{CP}$, $\mathcal{T}$ and $\mathcal{CPT}$, is the $K^0, \bar{K}^0$ (particle–antiparticle) base. In this representation

$$\begin{aligned} \mathcal{CP}\,|K_0^{\rm in}\rangle &= |\bar{K}_0^{\rm in}\rangle\,,\\ \mathcal{T}\,|K_0^{\rm in}\rangle &= \langle K_0^{\rm out}|\,,\\ \mathcal{CPT}\,|K_0^{\rm in}\rangle &= \langle \bar{K}_0^{\rm out}|\,. \end{aligned} \tag{11}$$

Without loss of generality, we can express the physical incoming states in terms of $|K_0^{\rm in}\rangle$ and $|\bar{K}_0^{\rm in}\rangle$ as:

$$\begin{aligned} |K_S^{\rm in}\rangle &= \frac{1}{N}\left((1+\alpha)|K_0^{\rm in}\rangle + (1-\alpha)|\bar{K}_0^{\rm in}\rangle\right)\,,\\ |K_L^{\rm in}\rangle &= \frac{1}{N}\left((1+\beta)|K_0^{\rm in}\rangle - (1-\beta)|\bar{K}_0^{\rm in}\rangle\right)\,, \end{aligned} \tag{12}$$

where α and β are complex variables associated with $\mathcal{CP}$, $\mathcal{T}$ and $\mathcal{CPT}$-violation (usually denoted by ϵ_S and ϵ_L, respectively), and N a normalization factor. Then, the respective equations for the outgoing states are not independent, but are determined by the orthogonality conditions for the physical states[3]

$$\begin{aligned} \langle K_S^{\rm out}| &= \frac{1}{\tilde{N}}\left((1-\beta)\langle K_0^{\rm out}| + (1+\beta)\langle \bar{K}_0^{\rm out}|\right)\,,\\ \langle K_L^{\rm out}| &= \frac{1}{\tilde{N}}\left((1-\alpha)\langle K_0^{\rm out}| - (1+\alpha)\langle \bar{K}_0^{\rm out}|\right)\,. \end{aligned} \tag{13}$$

where the normalisation factor N can always be chosen equal to $N = \sqrt{2(1-\alpha\beta)}$.[3] Using Eqs. (9), (12) and (13) the Hamiltonian can be expressed in the basis of $K^0, \bar{K}^0$ as

$$H = \frac{1}{2}\begin{pmatrix} (\lambda_{\rm L}+\lambda_{\rm S}) - \Delta\lambda\frac{\alpha-\beta}{1-\alpha\beta} & \Delta\lambda\frac{1+\alpha\beta}{1-\alpha\beta} + \Delta\lambda\frac{\alpha+\beta}{1-\alpha\beta} \\ \Delta\lambda\frac{1+\alpha\beta}{1-\alpha\beta} - \Delta\lambda\frac{\alpha+\beta}{1-\alpha\beta} & (\lambda_{\rm L}+\lambda_{\rm S}) + \Delta\lambda\frac{\alpha-\beta}{1-\alpha\beta} \end{pmatrix}\,, \tag{14}$$

where $\Delta\lambda = \lambda_L - \lambda_S$.

From Eq. (14), we can identify the $\mathcal{T}$-, $\mathcal{CP}$- and $\mathcal{CPT}$-violating parameters. Indeed:

- Under <u>$\mathcal{T}$-transformation</u>,

$$\langle K_0^{\rm out}|H|\bar{K}_0^{\rm in}\rangle \leftrightarrow \langle \bar{K}_0^{\rm out}|H|K_0^{\rm in}\rangle\,,$$

thus, the off-diagonal elements of H are interchanged. This indicates that the parameter $\epsilon \equiv (\alpha+\beta)/2$, which is related to the difference of the off-diagonal elements of H, measures the magnitude of the $\mathcal{T}$-violation.[a]

$$\frac{2}{N^2}\epsilon = \frac{\langle K_0^{\rm out}|H|\bar{K}_0^{\rm in}\rangle - \langle \bar{K}_0^{\rm out}|H|K_0^{\rm in}\rangle}{2\Delta\lambda}\,. \tag{15}$$

- Under <u>$\mathcal{CPT}$-transformation</u>,

$$\langle K_0^{\rm out}|H|K_0^{\rm in}\rangle \leftrightarrow \langle \bar{K}_0^{\rm out}|H|\bar{K}_0^{\rm in}\rangle\,,$$

and therefore, the parameter $\delta \equiv (\alpha-\beta)/2$, related to the difference of the diagonal elements of H, measures the magnitude of $\mathcal{CPT}$-violation.

$$\frac{2}{N^2}\delta = \frac{\langle \bar{K}_0^{\rm out}|H|\bar{K}_0^{\rm in}\rangle - \langle K_0^{\rm out}|H|K_0^{\rm in}\rangle}{2\Delta\lambda}\,. \tag{16}$$

- Under <u>$\mathcal{CP}$-transformation</u>,

$$\langle K_0^{\rm out}|H|K_0^{\rm in}\rangle \leftrightarrow \langle \bar{K}_0^{\rm out}|H|\bar{K}_0^{\rm in}\rangle\,,$$

and simultaneously

$$\langle K_0^{\rm out}|H|\bar{K}_0^{\rm in}\rangle \leftrightarrow \langle \bar{K}_0^{\rm out}|H|K_0^{\rm in}\rangle\,,$$

thus, *both* the diagonal and the off-diagonal elements of H are interchanged. Then, the parameters $\alpha = \epsilon+\delta$ and $\beta = \epsilon-\delta$, are the ones which measure the magnitude of $\mathcal{CP}$-violation in the decays of K_S and K_L, respectively.

3. CPLEAR Direct Measurement of Time-Reversibility

Having identified the $\mathcal{CP}$, $\mathcal{T}$ and $\mathcal{CPT}$-violating operations, one may construct asymmetries that measure discrete symmetry-violations. For instance, a time-reversal operation interchanges initial and final states, with identical positions and opposite velocities:

$$\mathcal{T}[\langle \bar{K}_0^{\rm out}(t_f)|K_0^{\rm in}(t_i)\rangle] = \langle K_0^{\rm out}(-t_i)|\bar{K}_0^{\rm in}(-t_f)\rangle\,. \tag{17}$$

Assuming time-translation invariance

$$\mathcal{T}[\langle \bar{K}_0^{\rm out}(t_f)|K_0^{\rm in}(t_i)\rangle] = \langle K_0^{\rm out}(t_f)|\bar{K}_0^{\rm in}(t_i)\rangle\,. \tag{18}$$

[a] $2/N^2 \approx 1$, in the linear approximation.

The time evolution from t_i to t_f implies that

$$\langle \bar{K}_0^{\rm out}(t_f)|K_0^{\rm in}(t_i)\rangle = \frac{1}{N^2}(1-\alpha)(1-\beta)(e^{-i\lambda_S\Delta t} - e^{-i\lambda_L\Delta t}), \tag{19}$$

$$\langle K_0^{\rm out}(t_f)|\bar{K}_0^{\rm in}(t_i)\rangle = \frac{1}{N^2}(1+\alpha)(1+\beta)(e^{-i\lambda_S\Delta t} - e^{-i\lambda_L\Delta t}). \tag{20}$$

Then, by definition, the magnitude of $\mathcal{T}$-violation is directly related to the Kabir asymmetry[5]

$$\begin{aligned} A_{\mathcal{T}} &= \frac{|\langle K_0^{\rm out}(t_f)|\bar{K}_0^{\rm in}(t_i)\rangle|^2 - |\langle \bar{K}_0^{\rm out}(t_f)|K_0^{\rm in}(t_i)\rangle|^2}{|\langle K_0^{\rm out}(t_f)|\bar{K}_0^{\rm in}(t_i)\rangle|^2 + |\langle \bar{K}_0^{\rm out}(t_f)|K_0^{\rm in}(t_i)\rangle|^2}, \\ &= \frac{|(1+\alpha)(1+\beta)|^2 - |(1-\alpha)(1-\beta)|^2}{|(1+\alpha)(1+\beta)|^2 + |(1-\alpha)(1-\beta)|^2} \approx 4{\rm Re}\,[\epsilon]\,, \end{aligned} \tag{21}$$

which is time-independent. Any non-zero value for $A_{\mathcal{T}}$ signals a direct measurement of $\mathcal{T}$-violation without any assumption about $\mathcal{CPT}$ invariance. Here, we should note that in linear order in ϵ and δ, the approximate equality

$$\langle K_S^{\rm in}|K_L^{\rm in}\rangle + \langle K_L^{\rm in}|K_S^{\rm in}\rangle \approx 4{\rm Re}\,[\epsilon]\,, \tag{22}$$

holds. This follows directly from the non-orthogonality of the adjoint states $\langle K_S^{\rm in}|$ and $\langle K_L^{\rm in}|$ that is manifest in the equations

$$\begin{aligned} \langle K_S^{\rm in}|K_S^{\rm in}\rangle &= \frac{1+|\alpha|^2}{|1-\alpha\beta|}, \qquad \langle K_L^{\rm in}|K_L^{\rm in}\rangle = \frac{1+|\beta|^2}{|1-\alpha\beta|}, \\ \langle K_S^{\rm in}|K_L^{\rm in}\rangle &= \frac{\alpha^*+\beta}{|1-\alpha\beta|}, \qquad \langle K_L^{\rm in}|K_S^{\rm in}\rangle = \frac{\alpha+\beta^*}{|1-\alpha\beta|}. \end{aligned} \tag{23}$$

However, although the time-reversal asymmetry can *in the linear approximation* be expressed in terms of only incoming states, the conceptual issue of reversing the time-arrow for any $\mathcal{T}$-violation measurement is unambiguous. For this reason, the CPLEAR collaboration searched for $\mathcal{T}$-violation through K^0–$\bar{K}^0$ oscillations, a process where initial and final states can be interchanged.

CPLEAR produces initial neutral kaons with defined strangeness from proton–antiproton annihilations at rest, via the reactions

$$\mathrm{p\bar{p}} \longrightarrow \begin{cases} K^-\pi^+K^0 \\ K^+\pi^-\bar{K}^0\,, \end{cases} \tag{24}$$

and tags the neutral kaon strangeness at the production time by the charge of the accompanying charged kaon. Since weak interactions do not conserve strangeness, the K^0 and $\bar{K}^0$ may subsequently transform into each other via oscillations with $\Delta S = 2$. The final strangeness of the neutral kaon is then tagged through the semi-leptonic decays

$$\begin{aligned} K^0 &\to e^+\pi^-\nu\,, \quad \bar{K}^0 \to e^-\pi^+\bar{\nu}\,, \\ K^0 &\to e^-\pi^+\bar{\nu}\,, \quad \bar{K}^0 \to e^+\pi^-\nu\,. \end{aligned} \tag{25}$$

Among them, the first two are characterized by $\Delta S = \Delta Q$ while the other two are characterized by $\Delta S = -\Delta Q$ and would therefore indicate either (i) explicit violations of the $\Delta S = \Delta Q$ rule, or (ii) oscillations between K^0 and $\bar{K}^0$ that even if $\Delta S = \Delta Q$ holds, would lead at a final state similar to (i) (with the "wrong-sign" leptons). The CPLEAR experimental asymmetry is given by

$$A_{\mathcal{T}}^{\rm exp} = \frac{\overline{R}_+(\Delta t) - R_-(\Delta t)}{\overline{R}_+(\Delta t) + R_-(\Delta t)},$$

with

$$\begin{aligned}\overline{R}_+(\Delta t) &= |\langle e^+\pi^-\nu(t_f)|\bar{K}_0^{\rm in}(t_i)\rangle \\ &\quad + \langle e^+\pi^-\nu(t_f)|K_0^{\rm in}(t_f)\rangle\langle K_0^{\rm out}(t_f)|\bar{K}_0^{\rm in}(t_i)\rangle|^2, \end{aligned} \tag{26}$$

$$\begin{aligned}R_-(\Delta t) &= |\langle e^-\pi^+\bar{\nu}(t_f)|K_0^{\rm in}(t_i)\rangle \\ &\quad + \langle e^-\pi^+\bar{\nu}(t_f)|\bar{K}_0^{\rm in}(t_f)\rangle\langle \bar{K}_0^{\rm out}(t_f)|K_0^{\rm in}(t_i)\rangle|^2. \end{aligned} \tag{27}$$

where the first term in each sum stands for (i) and the second for (ii) (thus containing the kaon oscillations multiplied by the matrix element for semileptonic decays through $\Delta S = \Delta Q$. The experimental asymmetry $A_{\mathcal{T}}^{\rm exp}$ therefore, besides ϵ, also contains the parameters x_- and y, where x_- measures $\Delta Q = -\Delta S$, while y stands for $\mathcal{CPT}$ violation in the decays.

$$A_{\mathcal{T}}^{\rm exp} = 4{\rm Re}\,[\epsilon] - 2{\rm Re}\,[x_-] - 2{\rm Re}\,[y]\,. \tag{28}$$

In the CPLEAR experiment, with the proper experimental normalizations, the measured asymptotic asymmetry is[6]:

$$\tilde{A}_{\mathcal{T}}^{\rm exp} = 4{\rm Re}\,[\epsilon] - 4{\rm Re}\,[x_-] - 4{\rm Re}\,[y]\,. \tag{29}$$

The average value of $\tilde{A}_{\mathcal{T}}^{\rm exp}$ was found to be $= (6.6 \pm 1.6) \times 10^{-3}$, which is to be compared to the recent CPLEAR measurement of $({\rm Re}\,[x_-] + {\rm Re}\,[y]) = (-2 \pm 3) \times 10^{-4}$, indicating that the measured asymmetry is related to the violation of time-reversal invariance. Conclusively, CPLEAR made a direct measurement of time-reversal violation, as we had already stated.[3] Similar arguments have been presented,[7] using the density matrix formalism for the description of the kaon system.

An interesting question to ask at this stage, is what information one could obtain from previous measurements plus unitarity.[8] Unitarity implies the relations

$$\begin{aligned}\langle K_L^{\rm in}|K_S^{\rm in}\rangle &= \Sigma_f \langle K_L^{\rm in}|f^{\rm in}\rangle\langle f^{\rm out}|K_S^{\rm in}\rangle\,, \\ \langle K_S^{\rm in}|K_L^{\rm in}\rangle &= \Sigma_f \langle K_S^{\rm in}|f^{\rm in}\rangle\langle f^{\rm out}|K_L\rangle\,, \end{aligned} \tag{30}$$

where f stands for *all* possible decay channels. Making the additional assumption that the final decay modes satisfy the relation $|f^{\rm in}\rangle = |f^{\rm out}\rangle \equiv \langle f^{\rm out}|^\dagger$ (which is equivalent to making use of $\mathcal{CPT}$-invariance of the final state interactions), it is possible to calculate the sum $\langle K_L^{\rm in}|K_S^{\rm in}\rangle + \langle K_S^{\rm in}|K_L^{\rm in}\rangle$, by *measuring only the branching ratios of kaon decays.* This is what can be done in K_L, K_S experiments, where only the *incoming kaon states* are used. (Note here, however, that in the next section we discuss a $\mathcal{T}$-odd asymmetry that can be measured in a single decay channel.) In the linear approximation, this sum is equal to $4\mathrm{Re}\,[\epsilon]$ (see Eq. (22)). However, this is an *indirect* determination of $\mathcal{T}$-violation, and would not have been possible if invisible decays were present. This is to be contrasted with the results of CPLEAR, which use only one out of the possible decaying channels, and does not rely at all on unitarity and or the knowledge of other decay channels than the one used in the analysis.[3]

4. $\mathcal{T}$-Odd Effects Versus $\mathcal{T}$-Reversal Violation

The KTeV experiment looks at the rare decay $K_L \to \pi^+\pi^-e^+e^-$ of which they have collected more than 2000 events. In particular, they measure the asymmetry in the differential cross-section, with respect to the angle ϕ between the pion and electron planes.[9] To give to the angle an unambiguous sign, they define ϕ according to

$$\sin\phi\cos\phi = (\mathbf{n}_e + \mathbf{n}_\pi)\cdot\hat{z}\,, \tag{31}$$

where $\mathbf{n}_e(\mathbf{n}_\pi)$ is the unit vector in the direction $\mathbf{p}_{e^-}\times\mathbf{p}_{e^+}$ $(\mathbf{p}_{\pi^-}\times\mathbf{p}_{\pi^+})$, and $\hat{z}$ is the unit vector in the direction of the sum of the two pion momenta.[9] A $\mathcal{T}$-odd observable is one that changes sign under the reversal of all incoming and outgoing three-momenta and polarizations. By construction, ϕ satisfies this property. The operation of $\mathcal{T}$-reversal, involves in addition to the operations mentioned, a flip of the arrow of time (i.e. exchanging initial and final states). The KTeV collaboration observes an asymmetry of nearly 14% about $\phi = 0$, thus identifying a $\mathcal{T}$-odd effect.

The important issue is to assess when such an effect can be interpreted as a direct measurement of $\mathcal{T}$-reversal violation, since nowhere have the initial and final states been interchanged.[10] The key ingredient that effectively allows one to invert the arrow of time in such a process, is the hypothesis of the unitarity of the S-matrix: $SS^\dagger = 1$. The S-matrix can be written in terms of the $\mathcal{T}$-matrix for a process $i \to f$, as

$$S_{if} = \delta_{if} + i\mathcal{T}_{if}\,, \tag{32}$$

where a *delta*-function for energy-momentum conservation is included in $\mathcal{T}_{if}$. Unitarity now implies:

$$\mathcal{T}^*_{fi} = \mathcal{T}_{if} - iA_{if}\,, \tag{33}$$

where $\mathcal{T}_{fi}$ is the amplitude for a process $f \to i$ (i.e. exchanging initial and final states), and A_{if} is the absorptive part of the $i \to f$ process:

$$A_{if} = \sum_k \mathcal{T}_{ik}\mathcal{T}^*_{fk}\,, \tag{34}$$

and the sum extends over all possible on-shell intermediate states. Taking the absolute square of (33):

$$|\mathcal{T}_{fi}|^2 = |\mathcal{T}_{if}|^2 + 2\mathrm{Im}\,(A_{if}\mathcal{T}^*_{if}) + |A_{if}|^2 \,. \tag{35}$$

If $\tilde{\imath}$, $\tilde{f}$ denote the initial and final states with three-momenta and polarizations reversed, $\mathcal{T}$-reversal invariance would imply

$$|\mathcal{T}_{fi}|^2 = |\mathcal{T}_{\tilde{\imath}\tilde{f}}|^2 \,, \tag{36}$$

and from (35) we can construct

$$\begin{aligned} |\mathcal{T}_{if}|^2 - |\mathcal{T}_{\tilde{\imath}\tilde{f}}|^2 &= -2\mathrm{Im}\,(A_{if}\mathcal{T}^*_{if}) - |A_{if}|^2 \\ &\quad + (|\mathcal{T}_{fi}|^2 - |\mathcal{T}_{\tilde{\imath}\tilde{f}}|^2) \,. \end{aligned} \tag{37}$$

The left-hand side of (37) is precisely a $\mathcal{T}$-odd probability, for instance the one measured by KTeV. However on the right-hand side we have two contributions. The first contribution arises from the terms in the first line corresponding to final-state interactions (for instance the exchange of a photon between the π's and e's) which can affect the dependence on the angle ϕ and generate a $\mathcal{T}$-odd effect through $\mathcal{T}$-reversal conserving interactions. The other contribution, the last line of (37), is a genuine $\mathcal{T}$-reversal violating contribution. To identify a $\mathcal{T}$-odd effect with a violation of $\mathcal{T}$-reversal, it is thus necessary to estimate the effect of the final state interactions for the process concerned and to determine how big these contributions are with respect to the measured $\mathcal{T}$-odd effect. If these effects are small, then we can say that using unitarity (and $\mathcal{CPT}$ invariance of the final state interactions, which results in $\langle\pi^+\pi^-e^+e^-|^{\mathrm{out}} = (|\pi^+\pi^-e^+e^-\rangle^{\mathrm{in}})^\dagger$), we are effectively interchanging the roles of past and future and it is legitimate to identify the $\mathcal{T}$-odd effect with a measurement of $\mathcal{T}$-reversal violation.

5. Conclusions

In the light of the recent data by the CPLEAR and KTeV collaborations, we discuss violations of discrete symmetries in the neutral kaon system, with particular emphasis to $\mathcal{T}$-reversal violation versus $\mathcal{T}$-odd effects. Since decaying kaons correspond mathematically to a system with a non-Hermitean Hamiltonian, we use the dual space formalism, where the physical (decaying) incoming and outgoing states are distinct and dual of each other. This reflects the fact that the eigenstates of the full Hamiltonian may never oscillate to each other and have to be orthogonal at all times. The orthogonality conditions of the physical states, entirely determine the evolution of the kaon system. In this framework, we study both the asymmetries reported by CPLEAR and KTeV and conclude the following: CPLEAR, through K^0–$\bar{K}^0$ oscillations, effectively reverses the arrow of time and thus its measured asymmetry is directly related to the definition of $\mathcal{T}$-reversal. Having measured in the same experiment that additional effects which enter in the experimental asymmetry (arising by tagging the final kaon strangeness by semileptonic decays, i.e. violations of the $\Delta S = \Delta Q$ rule and $\mathcal{CPT}$-invariance in the decays) are small, it is concluded that CPLEAR indeed made the first direct measurement of $\mathcal{T}$-violation. Since the experiment uses only one out of the possible decaying channels, its results are also independent of any unitarity assumption, and the possible existence of invisible decay modes.

On the other hand, KTeV studies the decay $K_L \to \pi^+\pi^-e^+e^-$, which being an irreversible process measures $\mathcal{T}$-odd effects. These are not necessarily the same as $\mathcal{T}$-violating effects, since they reverse momenta and polarizations but not the time-arrow. It is straightforward to demonstrate that $\mathcal{T}$-odd and $\mathcal{T}$-violating effects are two different concepts. Non-vanishing $\mathcal{T}$-odd effects due to final state interactions, may arise even if unitarity and $\mathcal{T}$-invariance hold. However, since unitarity implies the inversion of the arrow of time, a $\mathcal{T}$-odd effect could be interpreted as time-reversal violation, provided $\mathcal{CPT}$-invariance of the final states holds and final state interactions are negligible.

Acknowledgments

We would like to thank A. de Rujula, for very illuminating discussions on $\mathcal{T}$-odd effects. The work of C.K. is supported by the TMR contract ERB-4061-PL-95-0789.

References

1. CPLEAR Collaboration, CERN-EP/98-153, *Phys. Lett.* **B444** (1998) 43.
2. J. Belz, for the KTeV Collaboration, hep-ex/9903025.
3. L.Alvarez-Gaumé, C. Kounnas, S. Lola and P. Pavlopoulos, hep-ph/9812326.
4. For a review, see N. W. Tanner and R. H. Dalitz, *Ann. Phys.* **171** (1986) 463; Some of the earlier references are: T. D. Lee and C. S. Wu, *Ann. Rev. Nucl. Sci.* (1966) 511; G. Sachs, *Phys. Rev.* **129** (1963) 2280; *Ann. Phys.* **22** (1963) 239; P. H. Eberhard, *Phys. Rev. Lett.* **16** (1966) 150; K. R. Schubert *et al.*, *Phys. Lett.* **B31** (1970) 662.
5. P. K. Kabir, *Phys. Rev.* **D2** (1970) 540.
6. CPLEAR Collaboration, A. Angelopoulos *et al.*, submitted to *Phys. Lett. B.*
7. J. Ellis and N.E. Mavromatos, hep-ph/9903386, contribution to the Festschrift for L. B. Okun, to appear in a special issue of *Physics Reports.*
8. S. Bell and J. Steinberger, in *Proc. Oxford Int. Conf. Elementary Particles*, Oxford, England, 1965, eds. R. G. Moorehouse *et al.*, p. 195.
9. L.M. Sehgal and M. Wanninger, *Phys. Rev.* **D46** (1992), 1035; *ibid.* 5209 (1992) (Erratum); P. Heiliger and L. M. Sehgal, *Phys. Rev.* **D48** (1993), 4146.
10. A. De Rujula, J. Kaplan and E. de Rafael, *Nucl. Phys.* **B35** (1971) 365; A. Bilal, E. Masso and A. De Rujula, *Nucl. Phys.* **B355** (1991) 549.

Shing-Tung Yau

GEOMETRY MOTIVATED BY PHYSICS

SHING-TUNG YAU
Department of Mathematics, Harvard University, USA
yau@math.harvard.edu

4 July 2003

It is a great honor to be invited for this important occasion. The first time I had met Prof. Yang was in the fall term of 1971 when he gave a series of public lectures at Stony Brook. I was deeply impressed by the intuitive power of physics as described by the great master. In the following years, I saw the impact of Yang–Mills theory on mathematics. It was exciting for a geometer to study the effect of physical intuition in our field and what we could do for physics in return. I spent the next thirty years exploring such an interaction. It was a fruitful research: from general relativity, to gauge theory and to string theory. The influence of Prof. Yang is very deep and I am grateful to him. In this note, I will explain such a personal experience. The choice of topic is therefore rather subjective.

The concept of space (or space-time) has evolved according to our ability in mathematics and our understanding of nature. In the ancient times, figures were constructed from line segments, planes and spheres (or quadrics like the parabola). Beautiful theorems in plane and solid geometry were proved by the Greeks. The most fundamental theorem is the Pythagorean theorem. Even modern geometry demands that this theorem holds infinitesimally. The fact that there are only five platonic solids had been a fascinating statement for mathematicians, philosophers and physicists (they are tetrahedron, octahedron, cube, icosahedron's, dodecahedron).

Although Archimedes had applied the concept of infinite process to geometry, it was not until the full development of Calculus (Newton, Leibniz) that we had the tools to study curved space. The foundation of calculus of variation due to Euler *et al.* also gave rise to many important geometric objects that are fundamental in modern geometry (e.g. minimal surfaces).

Euler also introduced the important concept of Euler number which is the foundation of all topological invariants. While geodesics and various geometric quantities were introduced on surfaces in three space, the very fundamental concept of intrinsic curvature was first introduced by Gauss. The product of two principle curvatures depends only on the first fundamental form and is independent of isometric deformation of surfaces. The product is called Gauss curvature, and its introduction is the birth of modern geometry, and it also inspired the famous work of Riemann. (Apparently Gauss was interested in geometry because he was asked to survey land.)

C. F. Gauss said that: "I am becoming more and more convinced that the necessity of our geometry cannot be proved, at least not by human reason nor for human reason. Perhaps in another life we will be able to obtain insight into the nature of space which is now unattainable. Until then we must place geometry not in the same class with arithmetic which is purely *a priori*, but with mechanics."

From this quotation of Gauss, we can see that he was deeply excited by what should be called space.

With the continuous development of mathematics and inputs from physics, we are facing a similar situation. Our concept of space may not be adequate.

Later, Riemann formally introduced the concept of abstract Riemann geometry which is free from being a subspace of Euclidean space.

Tensor Calculus was then developed on abstract space. Noncommutativity of covariant differentiation

$$D_{\frac{\partial}{\partial x}} D_{\frac{\partial}{\partial y}} \neq D_{\frac{\partial}{\partial y}} D_{\frac{\partial}{\partial x}}$$

gives rise to the concept of curvature of a connection.

In his approach to understand the multivalued analytical function, the concept of Riemann surface was introduced by Riemann. In order to make holomorphic function single valued, the concept of uniformization was introduced.

This development of Riemann surface has had a deep influence on modern geometry. It is a fundamental tool to understand surface in three space and was developed into the theory of complex manifold and projective geometry. They are the fundamental algebraic curves in higher dimensional space.

When Einstein formulated the unification theory of gravity with special relativity, the background mathematics was Riemannian geometry. His equation is

$$R_{ij} - \frac{R}{2} g_{ij} = T_{ij}$$

Ricci tensor is associated with matter while the full curvature tensor is related to full gravity behavior. This is a spectacular development. Geometry cannot be separated from physics anymore.

Any attempts to explicitly construct solutions of Einstein Equation, with physical boundary condition, end up with singularity. The spherical symmetric solution is found by Schwarzchild. It is the most important model for black hole. The basic question in general relativity is: Given a nonsingular set of generic data at time zero, what will the future space-time look like?

Penrose–Hawking proved that: If there is a trapped surface, singularity will develop. The theory of Penrose–Hawking is based on the technique of geometry. Only after this theory of "trapped surface", did we know singularity cannot be avoided. Penrose proposed the following fundamental question. For a generic nonsingular initial data set, the only possible future singularity is the type of black hole.

A closely related question is: can quantum gravity "cure" singularity of space-time. String theory does provide some hint that space-time may need a new concept.

However, we still need to answer the following: what is black hole singularity??

Let us now turn to another development in geometry. The development of Hamiltonian mechanics for celestial mechanics motivated the study of the behavior of geodesics for general Riemann manifold. Many important questions are being asked. For example, are there infinite number of closed geodesics on any compact manifold? For manifold with negative curvature, closed geodesics are closely related to the fundamental group of the manifold. The set of the lengths of closed geodesics is known to be related to the spectrum of the Laplacian. It has been very fruitful to study the spectrum of Laplacian from the point of view of semi-classical limit of quantum mechanics. The generalization to quantum field theory is remarkable.

An immediate generalization of geodesics is minimal surfaces. At the beginning, they are related to Plateau problems: surfaces that minimize areas with given boundaries.

The existence theory of minimal disk in a general Riemannian manifold was solved by Morrey whose method of regularity is fundamental for the theory of calculus of variations of two variables. When the minimal surface has no boundary, we can ask many questions similar to those for geodesics, however surfaces now have topology and they have to be taken into account in their calculations.

The question of how to count the number of minimal surfaces, their areas and their index has been useful in studying topology of three-dimensional manifolds. The study of stability of minimal surfaces gives rise to the study of black holes as developed by Schoen–Yau.

The calculus of variation developed by Morse, Morrey and others has been a powerful tool in geometry. Morse theory was used by R. Bott and S. Smale to solve an outstanding question in differential topology including periodicity theorem, and also to handle body decomposition theorems. Global theory of minimal surfaces has a lot of important applications. Some of these can be mentioned in the following. Sacks–Uhlenbeck demonstrated the existence of minimal sphere when the universal cover of the manifold is not contractible. Siu–Yau made use of this to prove the Frenkel conjecture that every compact Kähler manifold with positive curvature is CP^n. Schoen–Yau, Sacks–Uhlenbeck proved the existence of minimal surfaces with higher genus. Meeks–Yau, Meeks–Simon–Yau proved embeddedness of these surfaces. Thus questions on three-dimensional manifolds (e.g. Smith Conjecture) were solved when coupled with Thurston's theory as observed by C. Gordan. Schoen–Yau solved the positive mass conjecture based on minimal surface theory.

In another direction, the development of quantum mechanics strongly motivated the study of the spectrum of natural operators in geometry. The natural operators are Laplacian and Dirac operators. (For complex manifolds, they are $\overline{\partial}$ operators as well.) One of the great achievements was the proof of index theorem by Atiyah–Singer which allows us to count solutions of global differential equations when a certain obstruction vanishes.

A spectrum of operators give rise to topological invariants of manifolds: Zeta function was constructed through the heat kernel. The work of Ray–Singer on the determinant of the natural operators has become fundamental for later works on quantum field theory on manifolds.

Coupled with index theory is the Hodge theory and the Bochner vanishing theorem. When the curvature has a favorable sign, the vanishing theorem has been very powerful to establish existence theorems. However vanishing theorem does depend on Hodge theory which was motivated by two important subjects: fluid dynamics on two-dimensional surfaces and the Maxwell equations.

Hodge theory with coefficients in arbitrary bundle is one of the most powerful tools in modern mathematics. In the hands of Hodge, Kodaira, Hirzebruch, Atiyah, Deligne, Calabi, Griffiths, Schmid, Siu and others, the theory was used to study the structure of algebraic manifolds, especially for computing dimensions of holomorphic solutions of linear systems on complex manifolds. The very fundamental question of Hodge on the algebraic cycle is based on Hodge theory.

Maxwell equations can be considered as Abelian gauge theory in contrast to the Yang–Mills theory: the non-Abelian gauge theory. The works of E. Cartan on structure equations and Whitney on immersion theory had provided strong motivation to study the theory of Fiber bundle and Vector bundles. Important invariants like Stiefel–Whitney characteristic classes were needed to understand the topology of such bundles. In the forties, Pontryajin and Chern introduced their classes through connections over the bundles.

The curvature form representation of Pontryajin and Chern classes have been fundamental in topology and geometry. The idea of minimizing L^2-norm of curvature of connections on a given bundle did not appear in mathematics literature until 1954, when Yang–Mills published their work. At the same time, Calabi proposed to look at Kähler metric with minimal L^2-norm on its curvature. But the analysis was difficult in those days.

It was not until 1970s that the technique of nonlinear partial differential equations was more mature to deal with problems that arose from general relativity, gauge theory and string theory.

A deep understanding of linear theory is key to nonlinear theory. Good estimates for harmonic functions and eigenfunctions were developed by Cheng and myself. Applications to questions on affine geometry, Minkowski problem and maximal hypersurfaces were developed. Semi-linear elliptic equations were developed on the theory of conformal deformation of matrices by Yamabe, Trudinger, Aubin and Schoen.

The theory of quasi-linear equations was developed based on the understanding of regularity of minimal surfaces and harmonic maps. Works of Morrey, Nirenberg, De Giorgi, Federer–Fleiming, Almgren, Allard, Uhlenbeck, Simon, Schoen, Hamilton, and myself contributed to the theory in 1970s.

The development of Yang–Mills theory in mathematics started from the interaction of Yang with J. Simon who realized the Yang–Mills theory was about the theory of curvature of connections on fiber bundles.

In the mid-seventies, most efforts on the mathematics of Yang–Mills theory were concentrated on the self-dual equations over the four sphere. Note that the energy of Yang–Mills field can be shown to be not less than a suitable topological invariant of the bundle. The field is called self-dual or anti-self-dual if the inequality becomes equality. They played a fundamental role in physics. They are the so-called BPS states which are very stable states. The works of Atiyah–Singer–Hitchin initiated the application of index theorem to calculate the dimension of the moduli of self-dual Yang–Mills field. Then it was realized that Penrose twistor theory can be used to find all the solutions of the self-dual solutions of Yang–Mills field over the four sphere (Atiyah–Drinfeld–Hitchin–Manin).

In the late twenties, Karen Uhenback had developed the general elliptic theory of Yang–Mills theory. It is fundamental in understanding the compactification of the moduli space of Yang–Mills field.

C. Taubes found a remarkable way to construct self-dual solutions using singular perturbation. Based on these works, in 1983, Donaldson was able to study the global moduli space of self-dual solutions on a reasonable general four manifold. He then used the topology of the moduli space to create invariants of the four-dimensional manifolds. This remarkable development gave the first fundamental breakthrough in four-dimensional topology, giving a counter-example to Smale's theory of h-cobordism theorem in this dimension.

The Donaldson theory was later realized by Witten to be related to supersymmetric Yang–Mills theory, the theory developed by Seiberg and Witten (around 1995) was found to give topological invariants of four manifolds. It is believed to be closely related to the Donaldson invaraints. The Seiberg–Witten invariants were then applied to solve several important questions on geometry and topology. For example, algebraic surface of the general type cannot be diffeomorphic to rational surfaces. The Thom conjecture that the only embedded surface in a homology class represented by an algebraic curve must have genus not less than the genus of the algebraic curve, was proved by Kronheimer and Mrowka.

A very remarkable development was due to Taubes in relating the Seiberg–Witten invariants to the existence of Pseudo holomorphic curves in symplectic manifolds.

The supersymmetric Yang–Mills theory has many other important consequences in the modern development of string theory and in algebraic geometry including the understanding of counting algebraic curves in Calabi–Yau manifolds.

We shall also describe later the theory of Hamilton Yang–Mills connections on complex manifolds. The fact that anti-self-dual connection gives rise to holomorphic bundle was in fact observed by Prof. Yang. The theory developed by Donaldson–Uhlenbeck–Yau has important applications to algebraic geometry and string theory. We shall come back to this later.

In 1976, I solved the Calabi Conjecture on proving existence (and uniqueness) of Kähler–Einstein metrics. It includes the important case of Kähler Metric with zero Ricci curvature on manifolds with zero first Chern class.

To understand the significance of latter metrics, recall a most important class of solutions of Einstein equation that is Vacuua.

Traditionally, there are the following ways to find solutions:

1. Assume large groups of symmetries, either in Riemmanian or Lorentzian case. The most important example is Schwarzchild metric, which has maximal symmetrics.
2. Assume supersymmetries, i.e. assume there are parallel spinors.

The holomony group is then reduced to subgroups of $O(n)$. Besides locally symmetric space, the most important spaces that have special holomony groups are $SU(n)$, G_2 or Spin(7).

When the manifolds is Kähler (holomony group $= U(n)$), the problem reduces to solve the complex Monge–Ampere equation

$$\det\left(g_{ij} + \frac{\partial^2 u}{\partial z_i \partial \overline{z_j}}\right) = e^F \det(g_{ij})\,.$$

For a compact Kähler manifold, with first Chern class zero, there is a unique Ricci flat metric in each Kähler class. When there is cosmology constant α, the equation is

$$\det\left(g_{ij} + \frac{\partial^2 u}{\partial z_i \partial \overline{z_j}}\right) = e^{-\alpha u} e^F \det(g_{ij})$$

when $\alpha < 0$, it is much easier, when $\alpha > 0$, there are obstructions and is not completely understood. In the latter case, I conjectured twenty years ago its existence should come from the stability of the projective structure of the projective manifold. Simon Donaldson has recently made important progress on my conjecture.

My motivation to link stability of algebraic structures to solutions of elliptic equations comes from two sources:

1. In 1976, I proved the following Chern number in equations for algebraic surfaces with general type

$$3c_2(M) \geq c_1^2(M)$$

   The proof was based on the existence of Kähler–Einstein metric with negative constant. Miyaoka was able to prove the same inequality using the idea of Bogomolov, this idea was related to the theory of stability of bundle.
2. Based on the above mentioned work, I was motivated to study Yang–Mills connections over stable holomorphic bundles. This was proved by Donaldson for algebraic surfaces and Uhlenbeck–Yau for general Kähler manifolds.
3. In the theory of Ricci flow to change metric, the asymptotic behavior is clearly related to stability.

This theory of DUY is now an important piece of (2,0) theory in string models. Both the metric and the bundle theory contribute significantly to algebraic geometry and string theory. For string theory compactification, based on Kluza–Klein theory, both the Ricci feat metric and the Hermitian Yang–Mills contribute to the vacuua. For algebraic geometry, they are the building blocks for algebraic structures. It is remarkable that while we did not provide solutions of the metric and the connections in closed form, computation based on sophisticated algebraic geometry is possible.

Let me now describe ways to construct Calabi–Yau manifolds (Kähler manifolds with zero first Chern class).

(I gave a list of methods of construction in the Argone lab conference in 1983.)

1. Kummer Constructions, or more generally, orbifold construction.
2. Complete intersections of hypersurfaces in Fano manifolds, in particular, in the product of weighted projective spaces. The most important example was constructed by me in 1983.

$$\sum x_i^3 = 0, \sum y_i^3 = 0, \sum x_i y_i = 0$$

   in $CP^3 \times CP^3$.

   It admits an action of group of order 3, the Euler number is -6. This manifold is significant for building physical models as it gives three families of Fermions and it has nontrivial Wilson lines. Later Tian observed that my construction could be extended to more examples. But B. Greene observed they are deformations of the above manifold. Jun Li and I are able to deform the tangent bundle plus trivial bundle to stabilize $SU(4)$ bundle. This allows interpretation for Heterotic String Theory.
3. Hypersurface in Toric variety. The first such construction was due to Roan–Yau. Now it has become a standard construction. Batyrev observed mirror constructions of such manifolds in terms of duality of polyhedrons.
4. Branch cover construction. For example, take a quantic in CP^3. We can take branch cover of CP^3 along it to obtain quantic in CP^4. In general, any hypersurface can be associated with Calabi–Yau manifolds in high dimension.

Similar questions can be asked for noncompact complete Ricci flat manifolds. Vafa and his coauthors have considered this class of manifolds as local models and have been very successful in the computation of their instanton construction. How to understand the moduli space of such metrics? When the manifold is topologically finite, I conjectured in 1978 that M can be compactified to a Kähler manifold whose exceptional set in defined by an anticanonical divisor. The converse is basically true. When the anticanonical divisor is nonsingular, it is relatively easy and has been written up later. It was applied to construct semi-flat Ricci flat manifolds in my paper with Greene, Shapley and Vafa on the construction of Cosmic String.

When the anticanonical divisor is simple (normal crossing) the problem of parameterizing these metrics is very interesting. A very important case of our construction is that of algebraic manifold M with anticanonical divisor D to be nonsingular Calabi–Yau hypersurface (so that the normal bundle is trivial). The metric I constructed on $M \backslash D$ is cylindrical along D and can be glued along a tabular neighborhood of D to form a new CY manifold. Conversely for a large class of CY manifolds, we can pull it apart along a CY hypersurface. Gukov and I have been looking for the physics of such manifolds.

A very important understanding of Calabi–Yau manifold comes from the concept of mirror construction. It can be considered as a symmetry on the category of Calabi–Yau manifolds. This was suggested by Lerche–Vafa, and Dixon. Later, Greene–Plesser demonstrated its existence in the case of quantic. It realizes the duality between II_A and II_B theory from string theory. The duality allows one to compute the number of instantons arisen in II_A theory. These are the holomorphic curves in Calabi–Yau manifolds. The computation of the number of holomorphic curves in algebraic manifolds dates back to the nineteenth century. It was called enumerative geometry. Input from mirror symmetry settles this "kind" of old problem. It was initialized by Candalas *et al.* and finally solved by Liu–Lian–Yau (and Givental independently in case the manifold is more special). The work of Liu–Lian–Yau–Givental was a rigorous mathematical piece of work that settles the old problem. However, it did not give a true understanding of mirror symmetry. In 1995, Strominger–Yau–Zaslow proposed that Calabi–Yau manifold should have a foliation (singular) of special Lagrangian torus. The mirror manifold should be obtained by taking duality along the leaves of the foliation.

The SYZ construction is based on the newly developed M-theory. Therefore the geometric construction has support from the intuition of physics. The complicated question of singularities involved in such construction is expected to be solvable. The SYZ construction can be seen from the following diagram.

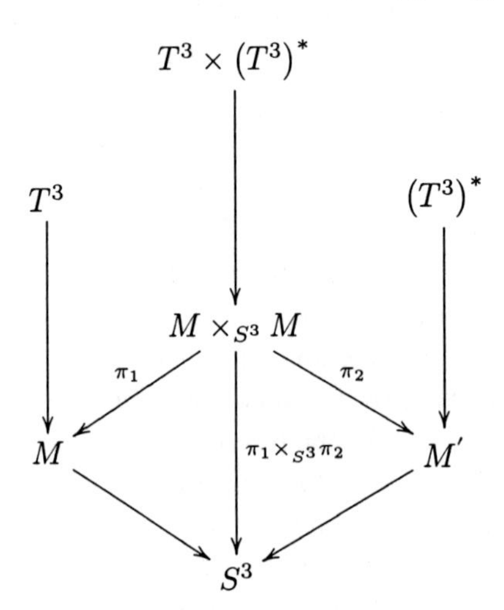

The above diagram allows us to transfer objects from M to M' and vice-versa. This generalized Fourier–Mukai transform was extensively studied in algebraic geometry.

From the SYZ construction, we know that special Lagrangian Cycles from M should move to stable holomorphic bundles over M and coupling should be preserved (Leung–Yau–Zaslow).

The construction should explain most of the questions in mirror symmetry. A very important one is how to map odd dimensional cohomology of a Calabi–Yau manifold to the even dimensional coholmology of its mirror.

I proposed the following diagram.

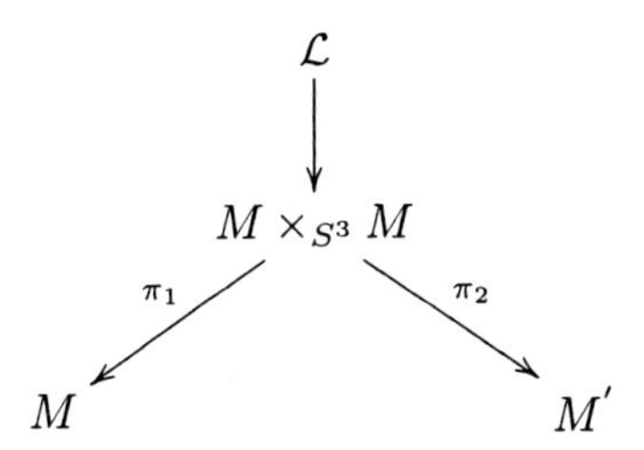

where $\mathcal{L}$ is the Poincare bundle defined fiberwisely on $T^3 \times (T^3)^*$ which produces a line bundle over the fiber product $M \times_{S^3} M'$. The map from odd cohomology of M to even cohomology of M' should be the composite of the following maps.

$$H^3(M) \longrightarrow H^3(M \times_{S^3} M')$$

$$\omega \longrightarrow \pi_1^* \omega$$

$$H^3(M \times_{S^3} M') \longrightarrow H^*(M \times_{S^3} M') \longrightarrow H^{even}(M')$$

$$\omega \longrightarrow \omega \wedge e^{c_1(\mathcal{L})} \longrightarrow (\pi_2)_* \left(\omega \wedge e^{c_1(\mathcal{L})}\right)$$

This map should exhibit the mirror symmetry in the cohomology level.

Many interesting questions arose in the SYZ construction. Most of the geometric quantities including complex structure and Ricci flat metrics will require quantum correction from disk instantons. (There is a background semi-flat Ricci flat metric given by Cosmic String construction.) How to compute such instantons are nontrivial. Works by C. C. Liu, Katz, Kefeng Liu, C. H. Liu, J. Zhou, myself and others are making progress over these important questions.

Around the same time as the development of SYZ, Kontzerich proposed homological mirror conjecture which suggested that derived category of M should be the same as the Fukaya category of M'. A lot of activities were initiated, especially by Bridgeland, Orlov, Kawamata, Thomas and others.

The concept of mirror symmetry was initiated by string theory. It does solve many outstanding questions in algebraic geometry that is otherwise difficult to understand. The ideas inspired by string theory in mathematics has been able to unify naturally many diverse concepts in mathematics. It helps to solve classical problems. It is inconceivable that nature will fool us by leading us so deeply into the core of mathematics.

While CY manifolds give a conformally invariant sigma model, a very interesting question is how to move from a general sigma model to a conformally invariant theory. This is provided by renormalization group flow.

In fact, twenty years ago, this was studied by Richard Hamilton. The equation is

$$\frac{dg_{ij}}{dt} = -2R_{ij} \, .$$

where g_{ij} is the metric and R_{ij} is its Ricci tensor.

The Hamilton equation has led to an important understanding of geometric structures on manifolds. I proposed to Hamilton to provide a proof of the geometrization program of Thurston based on this flow. He was able to carry out this program in much more detail. Some new impacts were given by Perelman recently.

In a closely related development, the dynamics of a surface moving by its mean curvature have shown some exciting development. Huisken and others have made important contributions. This will contribute to applied mathematics, geometry and physics.

The whole impact of geometry is moving dynamically, touching important parts of physics and engineering.

References

1. B. Andreas, G. Curio, D. H. Ruiperez and S.-T. Yau, *Fourier-Mukai Transform and Mirror Symmetry for D-Branes on Elliptic Calabi–Yau*, arXiv:math.AG/0012196.
2. B. Andreas, G. Curio, D. H. Ruiperez and S.-T. Yau, *Fiberwise T-Duality for D-Branes on Elliptic Calabi–Yau*, arXiv:math.AG/0101129.
3. M. F. Atiyah, N. J. Hitchin, V. G. Drinfeld and Yu. I. Manin, *Phys. Lett.* **A65**, 185 (1978).
4. P. Candelas, X. de la Ossa, P. Green and L. Parkes, *Nucl. Phys.* **B359**, 21 (1991).
5. S. K. Donaldson, *J. D. G.* **18**, 269 (1983).
6. S. K. Donaldson, *Proc. London Math. Soc.* **50**, 1 (1985).
7. A. Givental, *Equivariant Gromov–Witten Invariants, I. M. R. N* (1996), no. 13, 613–663.
8. B. Greene, A. Shapere, C. Vafa and S.-T. Yau, *Nucl. Phys.* **B337**, 1 (1990).
9. R. Hamilton, *J. D. G.* **17**, 255 (1982).
10. S. Katz and C. Liu, *A. T. M. P.* **5**, 1 (2001).
11. P. B. Kronheimer and T. S. Mrowka, *Math. Res. Lett.* **1**, 797 (1994).
12. B. Lian, K. Liu and S.-T. Yau, *Asian J. Math.* **1**, 729 (1997).
13. C. Liu, K. Liu and J. Zhou, *On a Proof of a Conjecture of Marino–Vafa on Hodge Integrals*, math.AG/0306257.
14. J. Li and S.-T. Yau, *Mathematical Aspects of String Theory* (World Scientific Publishing, Singapore, 1987), p. 560.
15. N. Leung, S.-T. Yau and E. Zaslow, *Winter School on Mirror Symmetry, Vector Bundles and Lagrangian Submanifolds* (Cambridge, MA, 1999), p. 209; *AMS/IP Stud. Adv. Math.* **23** (Amer. Math. Soc., Providence, RI, 2001).
16. S.-S. Roan and S.-T. Yau, *Acta Math. Sin.* (N.S.) **3**, 256 (1987).
17. R. Schoen and S.-T. Yau, *Phys. Rev. Lect.* **42**, 547 (1979).
18. R. Schoen and S.-T. Yau, *C. M. P.* **65**, 45 (1979).

19. A. Strominger, S.-T. Yau and E. Zaslow, *Nucl. Phys.* **B479**, 243 (1996).
20. C. Taubes, *J. D. G.* **17**, 139 (1982).
21. C. Taubes, *Seiberg Witten and Gromov Invariants for Symplectic 4-Manifolds*, First Intl. Press Lect. Ser. Vol. **2** (Internationl Press).
22. G. Tian and S.-T. Yau, *Mathemetical Aspects of String Theory* (World Scientific, Singapore, 1987), p. 543.
23. G. Tian and S.-T. Yau, *J. A. M. S.* **3**, 579 (1990).
24. K. Uhlenbeck and S.-T. Yau, *C. P. A. M.* **39**, 257 (1986).
25. E. Witten, *Math. Res. Lett.* **1**, 769 (1994).
26. S.-T. Yau, *C. P. A. M.* **31**, 339 (1978).

Bill Sutherland

BETHE'S ANSATZ: NOW & THEN

BILL SUTHERLAND
Physics Department, University of Utah,
Salt Lake City, UT 84112, USA

We present a simple lecture covering 35 years of research on exactly solvable one-dimensional quantum many-body systems — the Bethe's Ansatz — its generalizations and related developments. Although the selection of topics is my own, the work has been done by several others. Further, the contributions have built on one another in such a way that — to be honest — it is difficult to imagine one bit without the others. So, please see this talk as an appreciation of a long and fruitful history of cooperation, beginning with Bethe, and spanning generations.

1. Circa 1996

In 1966 when I began my thesis work with Professor C. N. Yang, there were only two or three exactly solved interacting quantum many-body problems — all one-dimensional. The solution to these went like this: Try as a solution a wavefunction — the so-called 'Bethe's Ansatz' — of the form

$$\Psi = \sum_{N!P's} A(P) e^{i\Sigma x_j k_{Pj}} . \tag{1}$$

This wavefunction is simply a summation over plane waves, with the particle momenta k_{Pj} being one of the $N!$ permutations P of the incoming momenta k_j. When it works, it works only if the coefficients are related by

$$A(\cdots, k', k, \cdots) = e^{-i\theta(k,k')} A(\cdots, k, k', \cdots) , \tag{2}$$

where θ is the 2-body phase shift, since N-body scattering must include 2-body scattering. Finally, applying periodic boundary conditions, the k's are determined by

$$kL = 2\pi I(k) + \sum_{k'(\neq k)} \theta(k, k') , \tag{3}$$

with $I(k)$ as the quantum numbers which determine the quantum state.

This then is the solution for those few solvable problems, and it can be used to determine the spectrum of the Hamiltonian.

2. Wavefunctions

Although in 1966, we had an explicit wavefunction, it was very difficult to say much about the correlations in such a quantum state. So, let us take a different approach.

The following product form for a wavefunction is much simpler than the Bethe's Ansatz:

$$\Psi_0 = \prod_{N \geq j > i \geq 1} \psi^{\lambda}(x_j - x_i)\,. \tag{4}$$

In fact, it is so simple that we might actually be able to evalute correlations in some cases; after all, it is just the Boltzmann weight for a classical pair potential $-v(r)/kT = \lambda \log \psi(r)$!

So we play a game: What Hamiltonian has a (groundstate) wavefunction of product form; i.e. given by Eq. (4). The Rule of this game is that the interaction must be a sum of pair potentials; otherwise the game would be trivial and no fun at all. Let us write the (time-independent) Schrödinger equation ($\hbar = m = 1$) in the form

$$\frac{1}{2\Psi_0} \sum_{N \geq j \geq 1} \frac{\partial^2 \Psi_0}{\partial x_j^2} = \sum_{N \geq j > i \geq 1} v(x_j - x_i) - E_0\,. \tag{5}$$

Substituting our product form into this equation, the only difficulty is with the cross-terms; our Rule requires that they add up to give a sum of pair potentials. If it works for three particles, it works for N particles. Thus, if $x + y + z \equiv (x_1 - x_2) + (x_2 - x_3) + (x_3 - x_1) = 0$, then we must satisfy the equation

$$\frac{\psi'(x)}{\psi(x)}\frac{\psi'(y)}{\psi(y)} + \frac{\psi'(y)}{\psi(y)}\frac{\psi'(z)}{\psi(z)} + \frac{\psi'(z)}{\psi(z)}\frac{\psi'(x)}{\psi(x)} = f(x) + f(y) + f(z)\,. \tag{6}$$

This is a functional equation, and the most general solution is an elliptic function. (We will carry out the algebra explicitly in the following Sec. 3; it's not difficult.)

One particularly interesting case is the pair potential

$$v(r) = \frac{\pi^2 \lambda(\lambda - 1)}{L^2 \sin^2(\pi r/L)} \underset{L \to \infty}{\longrightarrow} \frac{\lambda(\lambda - 1)}{r^2}\,. \tag{7}$$

The corresponding ground state wavefunction is

$$\Psi_0 = \prod_{N \geq j > i \geq 1} \sin^{\lambda}(\pi(x_j - x_i)/L) \to \prod_{N \geq j > i \geq 1} (x_j - x_i)^{\lambda}\,. \tag{8}$$

This wavefunction makes an intimate connection with the distribution of eigenvalues of a random matrix, for which there exists a large body of beautiful results, ready for exploitation.

Subsequent generalizations of this approach have been particularly fruitful.

3. Example: a Harmonic Trap

For the fun of it, let us carry out explicit calculations for a very simple example — the inverse-square pair potential. We begin, however, with the time-dependent Schrödinger equation

$$\frac{1}{2\Psi} \sum_{N \geq j \geq 1} \frac{\partial^2 \Psi}{\partial x_j^2} = V - \frac{i}{\Psi}\frac{\partial \Psi}{\partial t}\,. \tag{9}$$

We recover the time-independent Schrödinger equation, as usual, with

$$\Psi(x,t) = e^{-itE_0}\Psi_0(x)\,. \tag{10}$$

Now try the following product form for the wavefunction:

$$\Psi_0 = C \cdot \prod_{N\geq j\geq 1} e^{-\omega x_j^2/2} \cdot \prod_{N\geq j>i\geq 1} |x_j - x_i|^\lambda\,. \tag{11}$$

Substitute this into the left side of Eq. (9), and we get a page of algebra, which goes like this:

$$\begin{aligned}
\frac{1}{2\Psi}\sum_{N\geq j\geq 1}\frac{\partial^2\Psi}{\partial x_j^2} &= \text{diagonal terms} + \text{cross terms}\\
&= \frac{1}{2}\sum_{N\geq j\geq 1}(\omega^2 x_j^2 - \omega) + \sum_{N\geq j>i\geq 1}\frac{\lambda(\lambda-1)}{(x_j-x_i)^2}\\
&\quad -\omega\lambda\sum_{\text{pairs}}\left(\frac{x_1}{x_1-x_2}+\frac{x_2}{x_2-x_1}\right)\\
&\quad +\lambda^2\sum_{\text{triples}}\left(\frac{1}{x_1-x_2}\cdot\frac{1}{x_1-x_3}+\frac{1}{x_2-x_3}\cdot\frac{1}{x_2-x_1}+\frac{1}{x_3-x_1}\cdot\frac{1}{x_3-x_2}\right)\\
&= \frac{\omega^2}{2}\sum_{N\geq j\geq 1}x_j^2 + \sum_{N\geq j>i\geq 1}\frac{\lambda(\lambda-1)}{(x_j-x_i)^2} - \frac{N\omega}{2}[1+\lambda(N-1)]\\
&= \text{1-body term} + \text{2-body term} + \text{constant}\,.
\end{aligned} \tag{12}$$

Thus, we see that we have solved a problem with the potential

$$\begin{aligned}
V &= \frac{K}{2}\sum_{N\geq j\geq 1}x_j^2 + \sum_{N\geq j>i\geq 1}\frac{\lambda(\lambda-1)}{(x_j-x_i)^2}\\
&= \text{harmonic trap} + \text{interaction}\,.
\end{aligned} \tag{13}$$

The groundstate energy is

$$E_0 = \frac{N\omega}{2}[1+\lambda(N-1)]\,. \tag{14}$$

However, we have also solved a time-dependent problem! Let us suppose that $C(t)$ and $\omega(t)$ in our product form for the wavefunction, Eq. (11), are functions of time. Then on the right-hand side of Eq. (9) we find

$$\frac{i}{\Psi}\sum_{N\geq j\geq 1}\frac{\partial\Psi}{\partial t} = \frac{i}{C}\frac{dC}{dt} - \frac{i}{2}\frac{d\omega}{dt}\sum_{N\geq j\geq 1}x_j^2\,. \tag{15}$$

Combining this with the page of algebra, we see that we have a solution for the time-dependent potential

$$V(t) = \frac{K(t)}{2} \sum_{N\geq j\geq 1} x_j^2 + \sum_{N\geq j>i\geq 1} \frac{\lambda(\lambda-1)}{(x_j-x_i)^2}$$

$$= \text{time-dependent harmonic trap} + \text{interaction}\,, \tag{16}$$

provided $\omega(t)$ satisfies the Riccati differential equation

$$K(t) = \omega^2 - i\frac{d\omega}{dt}\,. \tag{17}$$

With the substitution

$$\omega = \frac{i}{\psi}\frac{d\psi}{dt}\,, \tag{18}$$

Riccati's equation is equivalent to the linear differential equation

$$\frac{d^2\psi(t)}{dt^2} + K(t)\psi(t) = 0\,. \tag{19}$$

So, we have an exact time-dependent wavefunction for a fluid of particles (1-D) interacting by an inverse-square pair potential, confined by a (arbitrary) time-dependent harmonic potential well.

This leaves us with one final question: Is this particular solution at all interesting? The answer is yes. It is the solution that corresponds to the ground state before we begin our time-dependent 'experiment', and so it is exactly the solution we would want, given a choice.

4. The Asymptotic Bethe's Ansatz

If a one-dimensional N-body system is (completely) integrable, then there are many (N) local conservation laws, of the form

$$\begin{aligned} N &= \sum_{N\geq j\geq 1} 1\,, \\ P &= \sum_{N\geq j\geq 1} k_j\,, \\ E &\underset{t\to\pm\infty}{\longrightarrow} \frac{1}{2}\sum_{N\geq j\geq 1} k_j^2\,, \\ &\vdots \\ L_N &\underset{t\to\pm\infty}{\longrightarrow} \sum_{N\geq j\geq 1} k_j^N\,. \end{aligned} \tag{20}$$

(We assume the potential to be repulsive, so as to scatter the particles.) Because of these conservation laws, scattering can only rearrange the asymptotic momenta

k_j, meaning there is no diffraction, meaning that the asymptotic form of the wave-function is given by Bethe's Ansatz

$$\Psi \underset{t\to\infty}{\to} \sum_{N!P's} A(P) e^{i\Sigma x_j k_{Pj}} , \tag{21}$$

where

$$A(\cdots, k', k, \cdots) = e^{-i\theta(k,k')} A(\cdots, k, k', \cdots) , \tag{22}$$

and θ is the 2-body phase shift. This is the *Asymptotic Bethe's Ansatz.*

Then, we can take a particle of momentum k around a ring. Along the way, it meets all the other $N-1$ particles, for a total change of phase

$$e^{ikL} \cdot e^{-i\theta(k-k_1)} \cdots e^{-i\theta(k-k_N)} = 1 . \tag{23}$$

Taking the logarithm, we have

$$kL = 2\pi I(k) + \sum_{k'(\neq k)} \theta(k, k') . \tag{24}$$

Once we have determined the k's, we can then evaluate the spectrum from Eq. (20). And so the problem is solved, much as the problems prior to 1966 were solved.

But is it really solved? Can we get a solution for finite density, from N-body asymptotic scattering; i.e. from zero density?

The answer is: Yes! Below, we outline the argument, which should resonate with statistical mechanicians.

There are two parts to the problem. First, one must be able to make a virial expansion of the thermodynamics, including zero temperature — meaning the ground state properties. This is nothing more than an expansion in powers of the density. Second, one must be able to calculate the nth order term of this expansion using only the n-particle scattering data.

The virial expansion was first introduced by Kamerlingh Onnes in 1901 to represent the thermodynamic data on real fluids as a deviation from the ideal gas, by means of a systematic expansion in powers of the density. As Uhlenbeck and Ford explained in their *Lectures in Statistical Mechanics*, such an expansion "...was not only desperation, but it contained the insight that the successive deviations from the ideal gas law will give information about the interaction of the molecules in pairs, triples, etc. This has been confirmed by the theoretical derivation ... from the partition function ..., first given in all generality by Ursell and Mayer around 1930."

The systematic expansion by Mayer — referred to above — was for a classical system. In Mayer's expansion, the nth order term required an integral over the coordinates of exactly n particles, without the constraint of a container. This does involve an exchange of the infinite volume limit, with the limit of an infinite number of particles. This is only justified for the low density or gas phase of the system, when the virial series converges to an analytic function of the density up to the singularity (possibly essential) at the phase boundary. A plausible mathematical mechanism for the development of such singular behavior — a phase transition — from the relatively well-behaved partition function, was presented in the beautiful papers of C. N. Yang and T. D. Lee that I read the summer before I began my thesis work with Professor Yang. This was the first time I ever saw that a physicist might need to worry in what order to take limits!

At that time when I was a graduate student, there was much interest in proving the convergence of the virial expansion. Many people were improving the estimates of the radius of convergence of the virial series for more and more realistic potentials. Of course, no one would expect a very long-ranged interaction such as the Coulomb potential to have a convergent expansion; just proving the existence of the thermodynamic limit for the Coulomb potential was a major achievement. But for potentials that decay rapidly, either classical or quantum, experience had convinced almost everyone that the equilibrium thermodynamics would have a convergent virial expansion. This was in sharp contrast, for example, to the formal attempts to write a virial expansion for transport properties.

The terms in the Mayer expansion for a classical system, as I have said, are expressed as so-called cluster integrals over the coordinates of the particles. A corresponding quantum cluster expansion was soon developed by Kahn and Uhlenbeck. However, the cluster functions required in the expansion are awkward to calculate. Instead, there is a very elegant expression, due to Beth and Uhlenbeck, which gives the second virial coefficient in terms of the two-body partial phase shifts $\delta_\ell(\varepsilon)$ as:

$$b_2 - b_2^o = 2^{3/2} \sum_{\substack{\text{bound}\\\text{states}}} e^{-\varepsilon/kT} + \frac{2^{3/2}}{\pi kT} \int_0^\infty d\varepsilon e^{-\varepsilon/kT} \sum_{\ell=0}^{\infty} (2\ell+1)\delta_\ell(\varepsilon)\,. \tag{25}$$

This expression appeared in 1937, and has a great appeal, since it expresses the second virial coefficient in terms of the experimentally measurable partial phase shifts instead of the unknown pair potential. This led Dashen, Ma and Bernstein in 1969 to re-investigate an old question: Can one express the thermodynamics of a system directly in terms of the scattering data, i.e. the on-mass-shell S-matrix? By directly, it is meant something like the Beth and Uhlenbeck formula, not simply inverting the two-body scattering to find the potential and then solving the resulting many-body problem. To summarize their long and difficult papers, they find the answer to be: Yes, provided the system is in the gas phase, or equivalently, if the system can support scattering. But of course, this is the only situation in which the S-matrix exists!

So to summarize more than 60 years of rather rigorous physics: If you have a system that supports scattering, and if you have all the n-body scattering data for this system, and if you can construct the corresponding virial series from this scattering data, and if you can sum this series and then analytically continue the result, then you have the exact thermodynamics.

But this is exactly what a technique first introduced by C. N. Yang and C. P. Yang will do for you, given the asymptotic Bethe's Ansatz! For systems that support scattering and are integrable, as we have seen, the N-body scattering is just a succession of 2-body scatterings, which is built into the asymptotic Bethe's Ansatz. Since the virial expansion uses only scattering data, it does not care whether the Bethe's Ansatz is asymptotic or exact. The method of Yang and Yang gives the exact thermodynamics for Bethe's Ansatz wavefunctions, so the thermodynamics it produces must be the analytic continuation of the virial series. When we take the scattering states given by the asymptotic Bethe's Ansatz and impose periodic boundary conditions by putting the particles on a ring with an arbitrarily large but finite circumference, this is obviously just a trick to help us to perform the appropriate integrations over the states. Any other trick or boundary condition would give the same thermodynamics. Similarly, if the potential is not of finite range, we must make it obey the boundary conditions somehow, but this will not affect the bulk properties.

5. How Do We Show Integrability?

The previous section began with the statement "If a one-dimensional N-body system is (completely) integrable ...". What followed was a solution. But how do we demonstrate that a system is integrable? In general, there is no best way; it depends upon the system. Three techniques have proven useful.

The first technique is 'try it and see'. This is the method that was originally used by Bethe; hence it is the 'classic' Bethe's Ansatz. For this to work, the asymptotic region must be everywhere, and so the potential must be short-ranged. The δ-function potential, Heisenberg–Ising model and Hubbard model were all first solved by this technique.

The second technique that has proven useful is to exploit the so-called Yang–Baxter equation or consistency conditions. When the particles are not identical, or if they carry quantum numbers, then there is the possibility of transmission and reflection. One must then be careful of the order in which a series of 2-body collisions occur. Otherwise, there will be diffraction from the 3-body region. This requires that certain products of 2-body amplitudes match, as in Fig. 1. From such consistency conditions, one can construct problems with known solutions. Surely, the most illustrious example is Baxter's solution of the 8-vertex and XYZ models. Again, the interactions tend to be short-ranged.

In this lecture, we want to emphasize a third technique — the so-called Lax method — which has proven useful for long-ranged potentials, and which justifies the asymptotic Bethe's Ansatz.

If we were to consider a classical system of N particles interacting by a pair potential, then it has been shown by Moser and Calogero — adapting a method of Lax — that for certain potentials one can find two Hermitean $N \times N$ matrices L and A that obey the Lax equation $dL/dt = i(AL - LA)$ Thus L evolves by a unitary transformation generated by A, and hence $\det[L - \lambda I]$ is a constant of motion. Expanding the determinant in powers of λ, we find N integrals of motion

$L_j, j = 1, \ldots, N$:

$$\det[L - \lambda I] = \sum_{j=0}^{N} (-\lambda)^{N-j} L_j \,. \tag{26}$$

Further, these integrals can be shown to be in involution, and thus the classical system is integrable.

Calogero then demonstrated that if one replaces classical dynamical variables with the corresponding quantum mechanical operators, $\det[L - \lambda I]$ is still well-defined with no ordering ambiguity. The Heisenberg equation of motion for an operator L is $dL/dt = i[H, L]$, where $[H, L]$ now represents the quantum mechanical commutator of the operator H with the operator L, as opposed to the previous matrix commutator $AL - LA$ in the Lax equation. In the same paper, Calogero then showed that the quantum mechanical commutator $[H, \det[L - \lambda I]] = 0$. Thus, the L_j are still constants of motion. Finally, Calogero showed that $[\det[L - \lambda I], \det[L - \lambda' I]] = 0$, and thus the two operators can be simultaneously diagonalized. However, the proof given by Calogero is very abbreviated and difficult to follow; an alternate proof sufficient for our purposes has been found that is very simple, yet elegant. I learned it from Sriram Shastry.

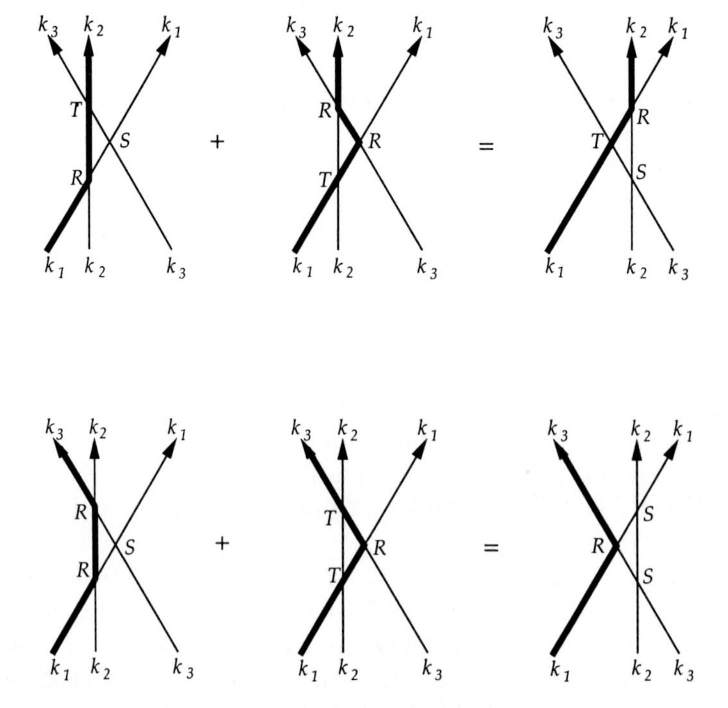

Fig. 1. Important identities for the matrix elements of the scattering operator implied by the consistency conditions. These are special cases of the Yang–Baxter equation.

We wish first of all to solve the quantum Lax equation

$$\frac{dL}{dt} = i[H, L] = i(AL - LA), \tag{27}$$

meaning: Find a Hamiltonian with pair potential $v(r)$, and two $N \times N$ Hermitean matrices with operator elements — the Lax A and L matrices — which obey the N^2 equations

$$HL_{jk} - L_{jk}H = \sum_{m=1}^{N} [A_{jm}L_{mk} - L_{jm}A_{mk}]. \tag{28}$$

From the quantum Lax equation, by itself, we can conclude nothing about the integrability of the systems. However, if there is a constant vector η such that $A\eta = \eta^{\dagger}A = 0$, then we can construct constants of motion by $L_n = \eta^{\dagger}L^n\eta$, since

$$\begin{aligned}[H, L_n] &= \eta^{\dagger}[H, L^n]\eta = \eta^{\dagger}\sum_{j=0}^{n-1}\{L^j[H, L]L^{n-1-j}\}\eta \\ &= \eta^{\dagger}\sum_{j=0}^{n-1}\{L^j(AL - LA)L^{n-1-j}\}\eta = \eta^{\dagger}\{AL^{n-1} - L^{n-1}A\}\eta = 0. \end{aligned} \tag{29}$$

Now from Jacobi's relation for commutators,

$$[H, [L_m, L_n]] + [L_n, [H, L_m]] + [L_m, [L_n, H]] = 0, \tag{30}$$

we see that the commutator of the constants of motion $[L_m, L_n]$ is also a constant of motion. However, looking at the explicit form of the Lax matrix L, we determine that asymptotically the Lax L matrix depends only on the momenta, and the constants of motion L_n are symmetric polynomials in the asymptotic momenta of degree n. Clearly then, the commutators of the constants of motion are asymptotically zero, but since they are also constants of motion they must vanish identically. We conclude then that the constants of motion commute with each other, and so can be simultaneously diagonalized. Thus the system is integrable.

The quantum Lax equation is again a functional equation, and once again, the most general solution is in terms of elliptic functions. Included in this family of soluble systems is the inverse-square potential.

6. Example: Those 'Pseudo'-Momenta

We now want to make a few comments about the parameters k_j, $j = 1, \ldots, N$ — the k's — which occur in the Bethe's Ansatz, Eqs. (1) and (21). These k's occur in plane waves, so they certainly seem like they should be the momenta of the particles — we even referred to them as the 'momenta' when we introduced the Bethe's Ansatz — but they are not. The k's are individually conserved, since they are integrals of motion, while the individual particle momenta are not conserved for an interacting system. For this reason, the k's are often called 'pseudomomenta'.

We also see this distinction from the expression for the energy of the system

$$E/L = \sum_k k^2/2L \underset{L\to\infty}{\to} \int \rho(k)k^2 dk/2\,. \tag{31}$$

Here $L\rho(k)$ is the distribution of k's. If the k's were 'real' momenta, then the total energy would be all kinetic energy! If the distribution of 'real' particle momentum p is given by $Ln(p)$, then the energy is

$$E/L = \int \rho(k)k^2 dk/2 = \int n(p)p^2 dp/2 + \text{potential energy}/L\,. \tag{32}$$

For an even more dramatic demonstration of the difference between 'real' and 'pseudo'-momenta, we can make an explicit calculation. We consider particles interacting by an inverse-square pair potential equation (7), so that the ground state wavefunction is given by Eq. (8). Now we need to say a little more about the statistics of the particles. Since the potential is strongly repulsive at the origin, we can impose statistics 'by force'. Using techniques from random matrix theory, we can then evaluate the momentum distribution $n(p)$ for the case $\lambda = 2$.

The results are shown in Fig. 2 for bosons and for fermions; they are very different. Bosons show a logarithmic singularity at zero momentum, reminiscent of the Bose–Einstein transition, while fermions have a vertical inflection point at the fermi momentum $p_f = \pi d$.

On the other hand, the 2-body phase shift for an inverse-square potential is constant — the scattering length is zero — so the fundamental equation for the asymptotic Bethe's Ansatz is easily solved for the ground state to give a distribution of k's

$$\rho(k) = \begin{cases} 1/2\pi\lambda, & |k| < k_f = \pi\lambda d; \\ 0, & |k| > k_f = \pi\lambda d\,. \end{cases} \tag{33}$$

This is independent of the statistics of the particles. (Note the interaction-dependent 'pseudo-fermi' momentum $k_f = \pi\lambda d$.) Thus, for $\lambda = 2$, the particles (bosons or fermions) in many respects behave as free fermions at twice the density, an example of so-called 'quasi'-statistics.

The asymptotic Bethe's Ansatz allows us to characterize the k's much more precisely than simply as 'not-real' momenta. Let us imagine we have an integrable system in equilibrium characterized by the usual thermodynamic parameters, say temperature and density. Then, suddenly we allow the system to expand freely into a vacuum, so that all integrals of motion — energy included — are conserved, keeping their equilibrium values. After a while, the particles have expanded so they no longer interact, and the momenta of the particles take their asymptotic values. But, referring back to the expressions of Eq. (20) for the integrals of motion, we see that the asymptotic momenta of the freely expanding particles must then be the k's of the asymptotic Bethe's Ansatz. The distribution of k's — our $\rho(k)$ above — can then be determined by time-of-flight measurements in this ballistic

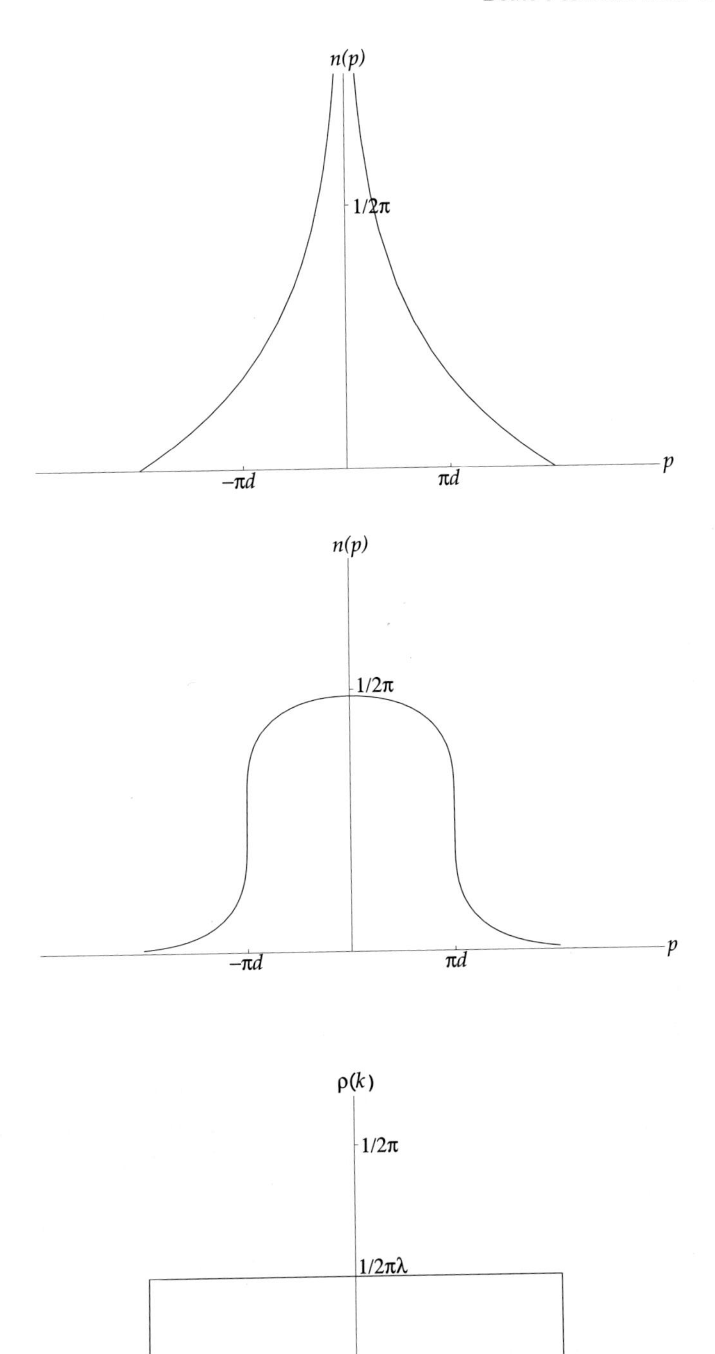

Fig. 2. The momentum distribution for interacting bosons (top) and fermions (middle), contrasted with the distribution of asymptotic momenta (bottom), which is independent of statistics.

expansion experiment, much as is done in the real Bose–Einstein experiments on cold atomic vapors.

In summary, the k's in the asymptotic Bethe's Ansatz are asymptotic momenta in a free-expansion experiment, as described above.

7. The Classical Limit

If we ask what these systems 'really look like', what we usually have in mind is a picture of point particles moving along trajectories — i.e. the classical limit. The systems with short-ranged potentials are in fact contact potentials, much like hard-spheres, confined to a line. However, the systems soluble by the Lax method include the long-ranged potential,

$$v(r) = \frac{\lambda(\lambda-1)\kappa^2}{\sinh^2(\kappa r)}. \tag{34}$$

This parameter λ is related to the usual potential strength g through $\lambda(\lambda-1) = gm/\hbar^2$, and so the classical limit corresponds to $\lambda \to +\infty$. Thus, this system has a nontrivial classical limit.

In the limit of vanishing density, the potential becomes the nearest-neighbor Toda lattice, a classical system about which much is known exactly. In particular, the particle branch of the (quantum) elementary excitations is identified with the classical soliton; the hole branch becomes a coherent beam of phonons.

At a more fundamental level, we can examine integrability itself, by integrating the differential equations for the classical system. We here present a portfolio of pictures of (one-dimensional) three-body classical scattering.

We have emphasized the concept of non-diffraction, and its relation to integrability; this is at the heart of our understanding of these systems. In classical mechanics an integrable system has a well-defined meaning: If the system is integrable, it has as many independent constants of motion as degrees of freedom. Therefore, if the mechanical system consists of particles scattering by a repulsive potential, then these constants of motion will fix the asymptotic momenta to be simply rearrangements of the incoming momenta. We claim that this is essentially the same as saying the scattering is non-diffractive. But diffraction is a concept from waves. What is the significance of diffraction in classical physics? Let us now explore classical scattering explicitly, with some examples, and some pictures.

We begin with 3-body scattering, taking as an example the Hamiltonian

$$H = (\dot{x}_1^2 + \dot{x}_2^2 + \dot{x}_3^2)/2 + \frac{1}{\sinh^2(x_2 - x_1)} + \frac{1}{\sinh^2(x_3 - x_2)} + \frac{1}{\sinh^2(x_1 - x_3)}. \tag{35}$$

In Fig. 3, we show three-body scattering with initial conditions chosen so that the particles come together at nearly (but not quite) the same place. We also show the straight-line motion of free particles with the same asymptotic conditions for comparison.

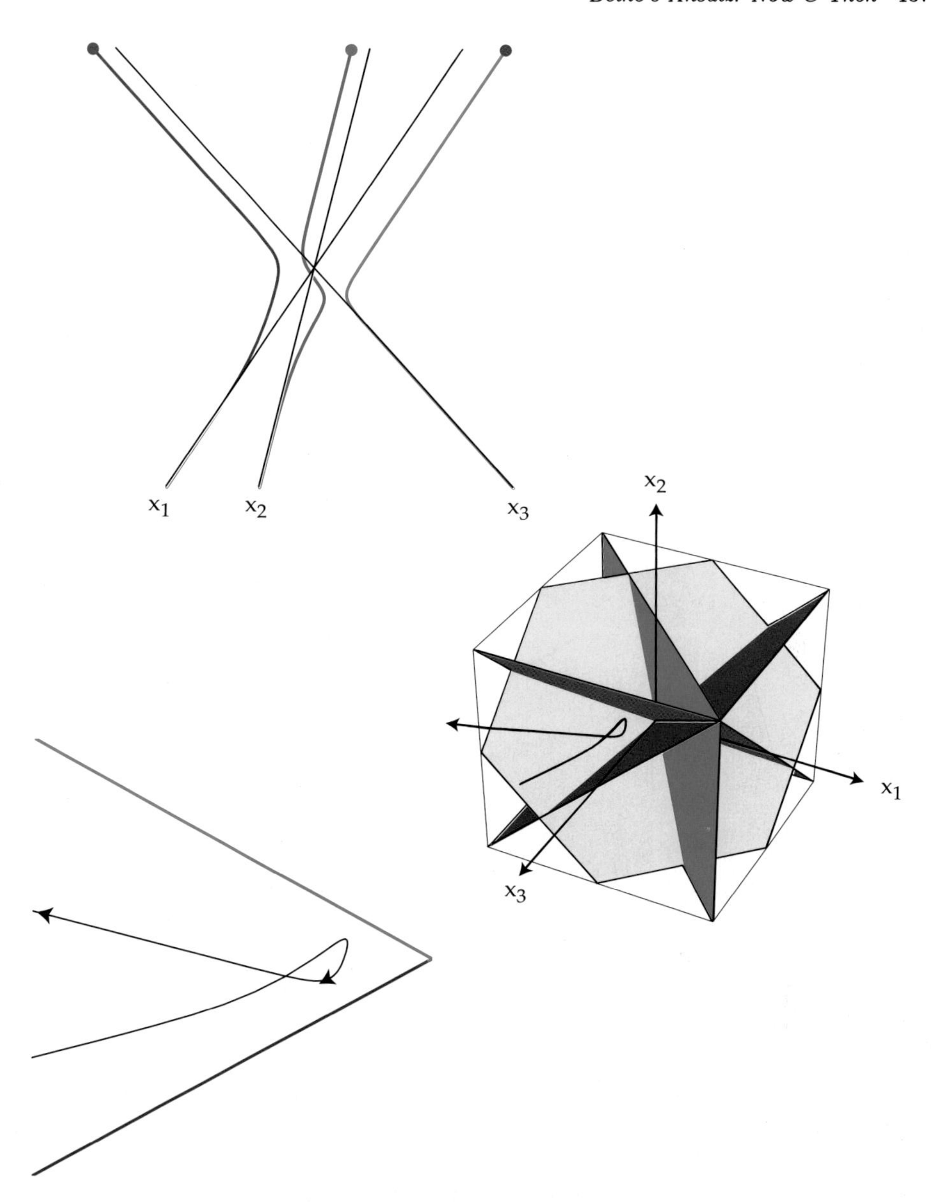

Fig. 3. Three pictures of 3-particle scattering: as three trajectories (top), as a point in three dimensions (middle), and in the center-of-mass frame (bottom).

A more useful plot is to treat the three coordinates x_1, x_2, x_3 as components of a single three-dimensional vector x. The mutual scattering of the three particles, which was represented by three individual trajectories, now becomes a single trajectory $x(t)$ in Fig. 3. Since the center of mass momentum is conserved, let us take $v_1+v_2+v_3 = 0$ and $x_1+x_2+x_3 = 0$. This is equivalent to projecting the motion onto the $x_1 + x_2 + x_3 = 0$ plane. It is quite useful for representing and understanding the

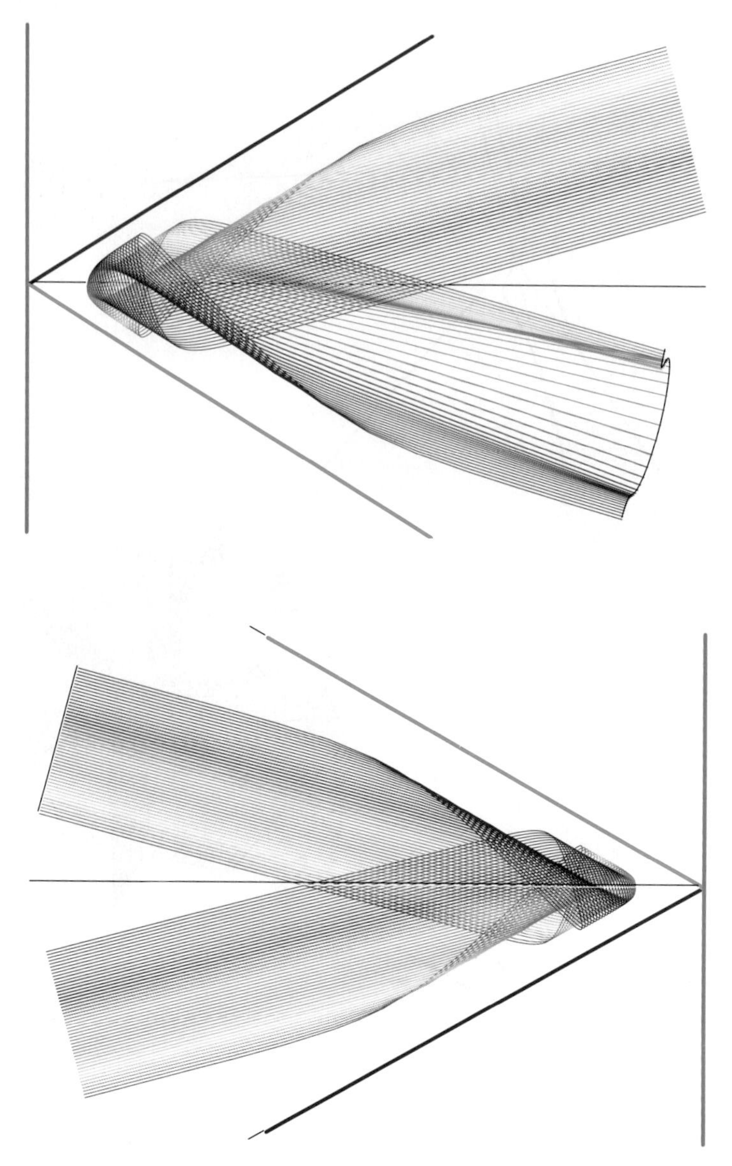

Fig. 4. Ray tracing in the center-of-mass frame of three particles, for a non-integrable system (top), and for an integrable system (bottom). Note the classical 'diffraction' in the non-integrable case.

physical problem. With new coordinates $x = (x_1 - x_3)/\sqrt{2}, y = (2x_2 - x_1 - x_3)/\sqrt{6}$, the three-body scattering in the center-of-mass frame also appears in Fig. 3. The heavy lines are the $x_1 = x_2$ and $x_2 = x_3$ planes.

Quantum mechanics is often called wave mechanics, because the classical trajectories are obtained as rays of the wave function, and so classical mechanics is

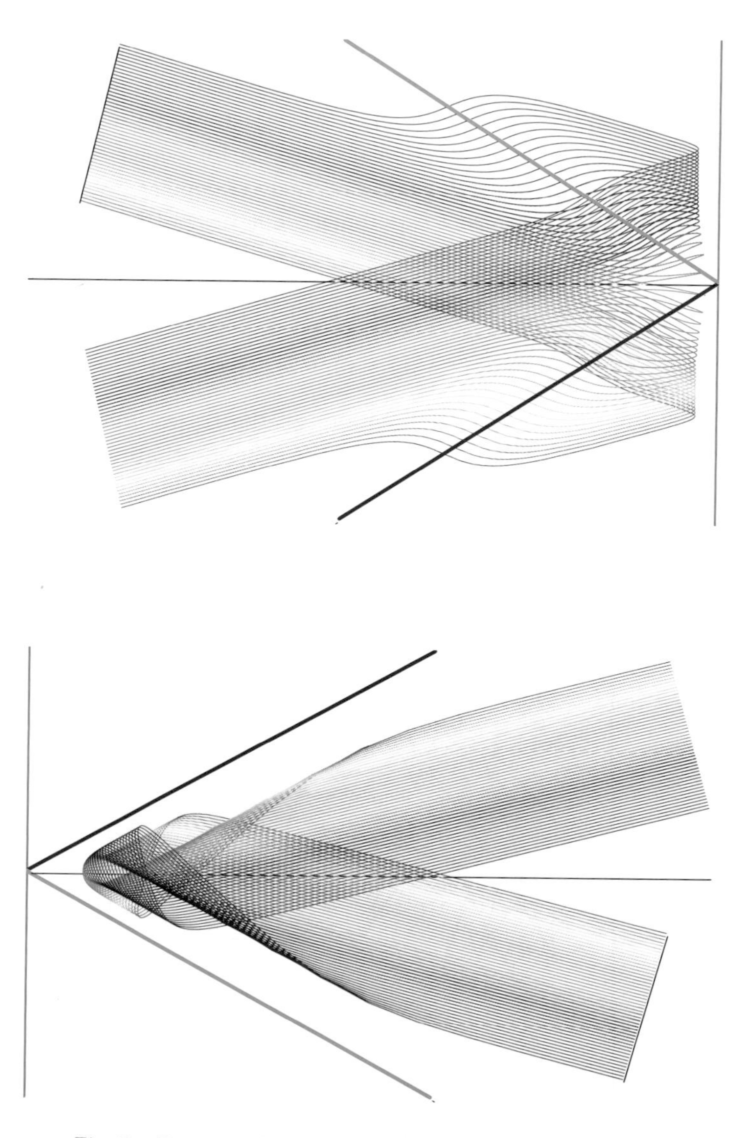

Fig. 5. Two more integrable cases, as described in the text.

then analogous to geometrical optics. Thus, we now take a number of trajectories, each with slightly different initial conditions, but all with the same asymptotic momenta — a bundle of rays. It is analogous to a monochromatic light beam. These trajectories then appear as a 'beam' in Fig. 4, where we have integrated the trajectories for the same amount of time, showing their final position by the heavy 'wave front'. We note the remarkable fact that the beam stays collimated, and the front is straight. This is exactly the case for specular reflection from mirrors due to point interactions. It is a property peculiar to an integrable system.

This is certainly not the generic situation, because usually we only have the conservation laws of energy and momentum, which are not enough to determine completely the asymptotic momenta. Thus, for trajectories for which three particles simultaneously interact, as opposed to trajectories for which the scattering is a sequence of three 2-body scatterings, the ray may emerge at any angle. In Fig. 4, we also show the same scattering experiment, but instead for the non-integrable potential

$$\begin{aligned}&\cosh^{-2}(x_2 - x_1) + \cosh^{-2}(x_3 - x_2) + \cosh^{-2}(x_3 - x_1)\\&\quad = \cosh^{-2}(\sqrt{2}x) + \cosh^{-2}((\sqrt{3}y + x)/\sqrt{2}) + \cosh^{-2}((\sqrt{3}y - x)/\sqrt{2})\,. \qquad (36)\end{aligned}$$

We clearly see the spreading, or 'diffraction', of the trajectories from the 3-body region. We once again emphasize that this is the generic situation, and so this single

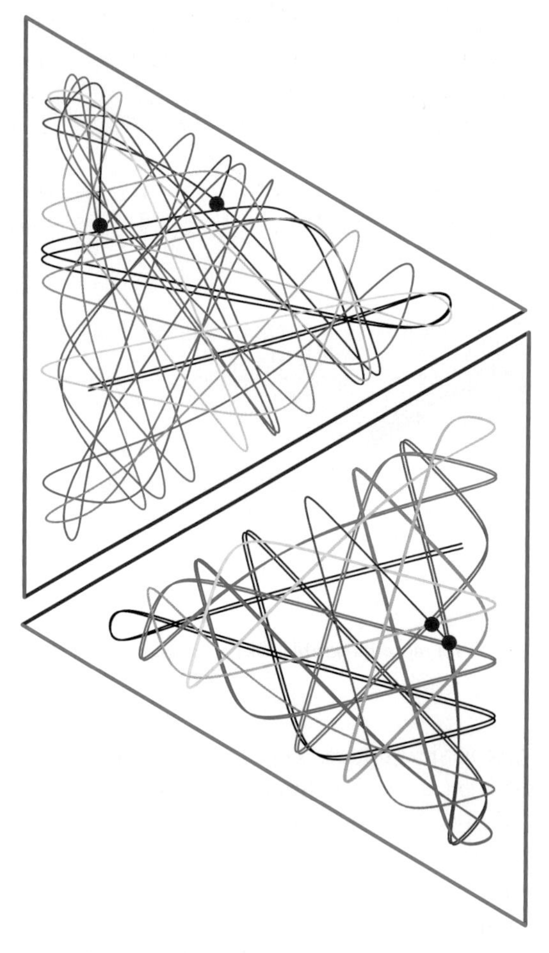

Fig. 6. Two trajectories with nearly identical initial conditions, for three particles on a ring, first for a non-integrable potential (top), and then for an integrable potential (bottom).

counter example serves as a (numerical) proof that the 3-body problem with the potential of Eq. (36) is not integrable.

On the other hand, if we modify the potential slightly to

$$\begin{aligned}\cosh^{-2}(x_2 - x_1) + \cosh^{-2}(x_3 - x_2) - \sinh^{-2}(x_3 - x_1) \\ = -\sinh^{-2}(\sqrt{2}x) + \cosh^{-2}((\sqrt{3}y + x)/\sqrt{2}) + \cosh^{-2}((\sqrt{3}y - x)/\sqrt{2}), \quad (37)\end{aligned}$$

i.e. if we change the interaction between the first and the last particles, making it attractive, then the beam scatters as in Fig. 5. (We change the sign of the potential in the two experiments.) This single case can never prove integrability, but it certainly makes us suspicious. (In fact, this system is integrable, as first shown by Calogero.)

The diffraction of the beam in the non-integrable, generic case will lead to a sensitive dependence on initial conditions, or chaotic behavior, and eventually causes the thermalization of the system. This is seen in Fig. 6 where we show two trajectories for the non-integrable periodic inverse-cosh-square potential — the periodic version of Eq. (36), corresponding to three particles on a ring. These two trajectories begin quite near, yet are soon completely uncorrelated. Contrast this with the same experiment for the integrable periodic inverse-sinh-square potential. (These periodic potentials are elliptic functions.) The particles soon play a game of 'follow-the-leader'. Thus, the differences between an integrable and a non-integrable system have been made quite graphic.

Akira Tonomura

THE MICROSCOPIC WORLD UNVEILED BY ELECTRON WAVES*

AKIRA TONOMURA

Advanced Research Laboratory, Hitachi, Ltd., Hatoyama, Saitama 350-0395, Japan
E-mail: tonomura@harl.hitachi.co.jp

Received 9 February 2000

Electron interference has been investigated at an industrial laboratory of Hitachi, Ltd. for the past 30 years thanks to the encouragement and support of C. N. Yang, who retired from the State University of New York in 1999. This paper reports here fundamentals and applications of electron interference experiments we have carried out including our relation to C. N. Yang, and is based on a talk made in honor of him at the "Symmetries and Reflections" Symposium held at Stony Brook in May 1999.

1. Introduction

In 1979, after 10 years of technical struggles, we developed a bright and coherent 80-kV field-emission electron source[1] for electron holography in transmission electron microscopes. The field-emission electron gun, developed by A. V. Crewe *et al.*, in 1968 was used not only to increase the resolution of a scanning electron microscope (SEM)[2] but also to observe individual heavy atoms with the newly developed scanning transmission electron microscope (STEM).[3] Attempts were naturally made to use this beam as a coherent wave, but they initially did not go well for the following reason. If a wide region of spatial beam-coherence is to be obtained, the divergence angle of an electron beam has to be kept within 10^{-8} rad. This degree of parallelity, however, was easily destroyed by the mechanical vibration of the electron source and by the deflection due to stray AC magnetic fields. It therefore took many years to get over this and other technical problems.

There was a reason we did not want to give up the development of a coherent electron wave. We demonstrated in 1969 that the holography devised theoretically by D. Gabor[4] was possible even with electrons,[5] but we soon realized that coherent electron optics,[6] including electron holography, would not be practically feasible without a much brighter electron beam. Soon after developing our field-emission electron beam, we began to implement our ideas dreamt during the fruitless period of the apparatus development. In 1980 we found a beautiful relation that when

*Dedicated to Prof. C. N. Yang on the occasion of his retirement.

we used electron holography to obtain an electron interference micrograph of a magnetic object, the contour fringes in the micrograph directly indicate projected magnetic lines of force in h/e flux units.[7]

While we were absorbed in directly observing microscopic magnetic domain structures in ferromagnetic thin films, there arose in fundamental quantum mechanics a controversy[8] over the existence of the Aharonov–Bohm effect.[9] We were not indifferent to this controversy because the Aharonov–Bohm (AB) effect provides the most fundamental principle underlying electron interference microscopy of magnetic objects. To put it more concretely, an interference micrograph of a magnetic object cannot be interpreted consistently without the AB effect. Furthermore, T. T. Wu and C. N. Yang[10] gave us a deep insight into the significance of the AB effect in relation to the gauge theory. In this theory, gauge fields, or vector potentials, are regarded as the most fundamental physical quantity.

As C. N. Yang often mentioned in his talks,[11] vector potentials were first introduced into the theory of electromagnetism as a physical entity by J. C. Maxwell: Maxwell identified the electro-tonic state Faraday regarded as a quantity more fundamental than electric fields or magnetic fields to be an "electromagnetic momentum" $\boldsymbol{A}$ describable by vector potentials. In fact, Maxwell showed that a magnetic field $\boldsymbol{B}$ is produced when there is a rotation in the spatial distribution of $\boldsymbol{A}$ ($\boldsymbol{B} = \text{rot}\ \boldsymbol{A}$) and that an electric field $\boldsymbol{E}$ is produced when $\boldsymbol{A}$ changes with time $\left(\boldsymbol{E} = -\frac{\partial \boldsymbol{A}}{\partial t}\right)$. The Maxwell equations, however, were not easy to understand and were not readily accepted. The Maxwell equations received wide recognition only after 30 years of effort by O. Heaviside and H. R. Hertz, who eliminated vector potentials from the Maxwell equations as "rubbish." In 1954, though vector potentials were revived as a fundamental physical quantity in the Yang–Mills theory,[12] which was the first step toward the construction of a unified theory of fundamental forces that was based on the gauge principle. The AB effect provides direct evidence for the gauge principle and the physical reality of vector potentials.

We wanted to establish a firm experimental foundation for the AB effect by using our newly developed bright electron beam and the most advanced photolithography techniques to actually carry out an AB-effect experiment in the perfect form that had been only a thought experiment. I wrote a letter to C. N. Yang, a complete stranger at that time, asking whether he thought such an experimental verification would be worthwhile. This letter marked the beginning of our relationship with C. N. Yang, one which continues till this day.

2. The Aharonov–Bohm Effect

In response to my letter concerning the experimental test of the AB effect, Yang was kind enough to visit the Hitachi Central Research Laboratory, located at Kokubunji in the suburbs of Tokyo, in 1981 to discuss our experimental plan. This was possible because just after my letter reached him he happened to be visiting the University of Tokyo and staying there for a month. Since this was the longest stay he made in Japan, it was really a fortunate chance for us.

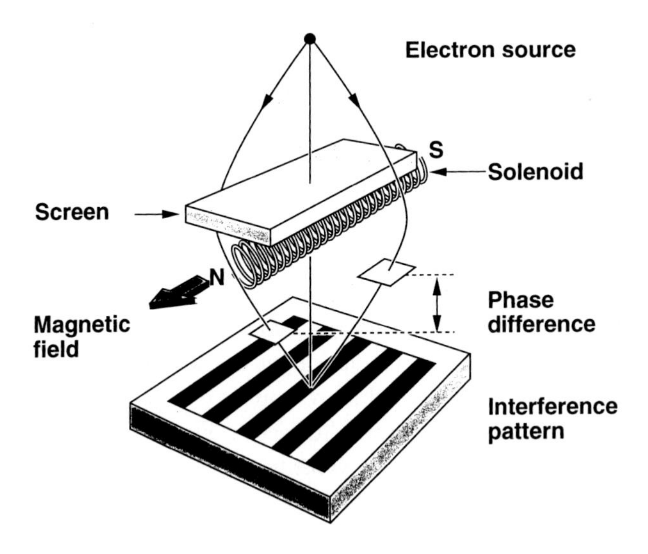

Fig. 1. The magnetic Aharonov–Bohm effect.

The AB effect in question was predicted by Y. Aharonov and D. Bohm in 1959.[9] They devised two kinds of experiments, one electric and the other magnetic, and predicted that in both experiments electron beams passing through field-free regions would be influenced observably. It is the magnetic AB effect that is now often referred to simply as the AB effect, since the experiment in which the electric one would be manifest is technically difficult to carry out.

The magnetic AB effect is shown in Fig. 1. When a current is applied to a solenoid, it becomes an electromagnet: magnetic poles are produced at both ends of the solenoid, thus generating magnetic fields inside and outside the solenoid. If the solenoid were infinitely long, however, there would be no magnetic fields outside the solenoid because the two poles go in the infinitely far-off distance.

When two electron beams from a point source pass on both sides of the solenoid and then overlap, an interference pattern is produced. Solving the Schrödinger equation, Aharonov and Bohm obtained a surprising result: the interference fringes are displaced by the magnetic flux passing through the solenoid even when the magnetic flux is confined within the solenoid and the electron beams never touch the magnetic flux. They attributed the relative phase shift $\Delta\phi$ between the two beams to the vector potentials surrounding the solenoid:

$$\Delta\phi = -\frac{e}{\hbar}\oint \boldsymbol{A}\, ds\,. \tag{1}$$

The vector potentials in a field-free region outside the solenoid cannot vanish in any gauge, since according to the Stokes' theorem the circulating components of vector potentials have to exist unless the magnetic flux in the solenoid vanishes. They asserted that, in quantum mechanics, vector potentials are physical quantities more fundamental than electric and magnetic fields.

Several experiments related to the AB effect were soon reported: Chambers,[13] for example, observed the interference fringes produced when two beams passed on either side of a pointed iron whisker magnetized in the direction of its length and found that the displacements of the fringes depended on the diameter of the iron whisker at the tapered region where magnetic flux passing through the whisker changed because of the flux leakage.

When the theory of gauge fields began to be regarded as the theory most likely to provide a unified explanation of fundamental forces, Bocchieri and Loinger[14] asserted in 1978 that the AB effect did not exist. The assertion brought about a controversy.[8] The experiments hitherto performed, they said, were not related to the AB effect because magnetic fields had leaked outside from the solenoids or magnets and thus influenced the electron beams directly. The controversy was limited to theoretical arguments and seemed to continue endlessly. We were convinced that the only way to obtain a conclusion would be to test the AB effect by carrying out conclusive experiments. The problem was how to fabricate a solenoid without any leakage magnetic flux. One cannot actually make an infinitely long solenoid. Is there any other way to exclude the possibility of leakage flux? — Yes, there is. Some people had already proposed the use of a toroidal solenoid.[15] With it, magnetic flux would be confined within the toroid. A toroidal sample, however, should be small than the spatial coherence length of an electron beam, say, a few tens of micrometers. Extremely tiny toroidal magnets have thus to be fabricated.

The experiment we devised for testing the AB effect seemed technically difficult but possible because we had the advantage of using advanced techniques at our industrial laboratory: the coherence region of an electron beam could be expanded by using a coherent electron beam just developed, leakage magnetic fields could be avoided by using a tiny toroidal magnet fabricated using most advanced photolithography techniques, and the leakage flux could be measured by electron interference microscopy.[6]

We showed such an experimental design to C. N. Yang when he visited our laboratory. He explained the essence and significance of the AB effect and suggested that we should quantitatively compare the phase shifts with the magnetic fluxes.

We were not only deeply moved by the trouble C. N. Yang took to visit our laboratory, but were also impressed with him because he explained difficult theories to laymen like us in simple and lucid terms. We became ardent admirers and dared to read the theoretical papers in which he describes the physical essence with sentences so plain and beautiful that even we experimentalists can understand them — or more exactly, can feel we understand. I still remember that although in my university days I made great efforts to learn the mathematics needed for understanding physics, I was often unable to reach the point where I understood and enjoyed physics. One of the reasons I joined industry just after getting my bachelor degree was that the mathematical barriers had robbed me of the fun brought to me by physics, though the main reason was that it was difficult for me to compete with smart classmates in physics.

Until I heard Dyson's banquet speech[16] at Prof. Yang's Retirement Symposium, I had believed that Prof. Yang's gift for explanation was due to his innate supreme ability and deep insight into physics. I noticed during this speech, however, that this was not the sole reason. According to Dyson, "Yang learned more physics from Fermi than from anybody else, and Fermi's way of thinking left an indelible impression in his mind." I am confident after reading the following writings in Prof. Yang's Selected Papers[17] that he must have made every effort to follow Fermi.

- "Fermi gave extremely lucid lectures... he started from the beginning, treated simple examples and avoided as much as possible "formalisms".... The simplicity was the result of careful preparation and of deliberate weighing of different alternatives of presentation.... We learned that *that* was physics. We learned that physics should not be a specialist's subject.

A beautiful example can be found in Wu and Yang's paper.[10]

- "Electromagnetism is thus the gauge-invariant manifestation of a nonintegrable phase factor."

We were delighted to know that all the electromagnetic phenomena, which take various aspects, are condensed in this one sentence. We were also encouraged by his following words,[12] which we read just before carrying out the AB-effect experiment.

- "The concept of an SU_2 gauge field was first discussed in 1954. In recent years many theorists, perhaps a majority, believe that SU_2 gauge fields do exist. However, so far there is no experimental proof of this theoretical idea, since conservation of isotopic spin only suggests, and does not require, the existence of an isotopic spin gauge field. A generalized Bohm–Aharonov experiment would be."

3. Experimental Tests of the AB Effect

We enthusiastically started our experiments on the AB effect soon after Yang's visit. We were able to collaborate with the group working on magnetic devices at our laboratory because our company's executives were convinced, as the famous Prof. Yang had visited our laboratory, that the confirmation experiment of the AB effect must be an important work. They were convinced even though the AB effect was not widely known at that time in Japan.

One year later we succeeded in fabricating tiny toroidal magnets [Fig. 2(a)]. The leakage flux, measured by electron-holographic interferometry,[6] was too small to influence the conclusions. Using them, we obtained results showing the existence of the AB effect.[18] That is, a relative phase shift between two electron beams passing through two field-free regions inside the hole and outside a toroidal magnet which had a closed magnetic circuit was detected [see Fig. 2(b)].

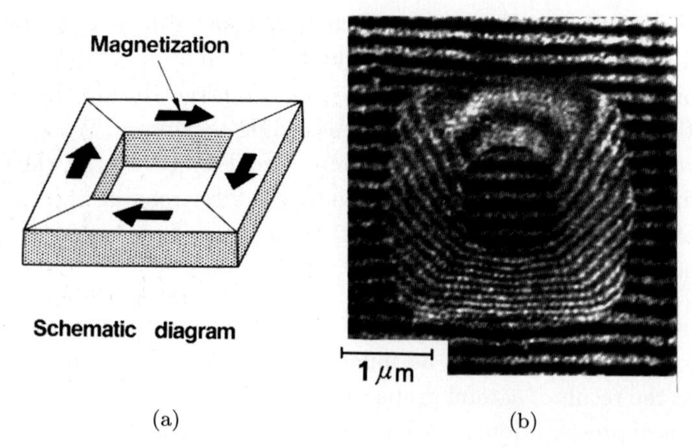

Fig. 2. The AB effect experiment using a transparent toroidal magnet made of permalloy. (a) Schematic; (b) electron interference pattern.

Soon after this result was reported, Bocchieri *et al.*[19] wrote a paper arguing against our interpretation. They asserted that the incident electron wave partly penetrated the magnet, and therefore the detected phase shift was not due to the AB effect. They demanded a better experiment, saying that the AB effect should be tested under the boundary conditions where the electron wave function completely vanished at the surface of the toroidal magnet.

In 1983 the first of a new series of international symposia discussing the fundamentals of quantum mechanics in the light of new technology (ISQM) was held at the Hitachi Central Research Laboratory. This symposium, based on the financial support provided by the Nishina Memorial Foundation, was originally intended to discuss the AB effect controversy. Nishina is the author of the famous "Klein–Nishina formula."

At the symposium, C. N. Yang proposed,[20] in his comments on my talk, a new experiment employing a toroidal magnet covered with a superconductor. His proposal reads as follows:

- "Your beautiful recent experiment has shown that, in a ring with minimum flux leakage, flux is not quantized. Can you also fabricate a small superconducting solenoid in an overall shape of a ring, with flux inside? If you can, then you can dramatically demonstrate flux quantization in the interference fringes' line-up matching."

We thought that Yang did not mean that our experiment was not convincing. I took his proposal as a suggestion to carry out not only a more conclusive experiment of the AB effect but also to directly demonstrate magnetic flux being quantized in a superconductor.

The experiment seemed to be much more difficult than the previous one: we had to fabricate a tiny and complicated sample — that is, a toroidal magnet completely

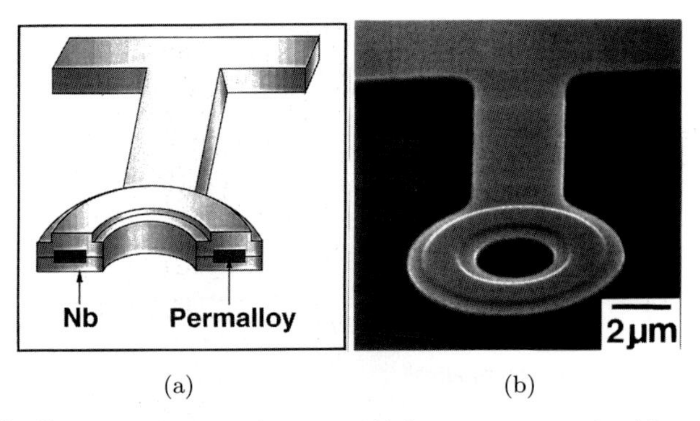

Fig. 3. The AB effect experiment using a toroidal magnet covered with a superconductor. (a) Schematic; (b) scanning electron micrograph.

covered with a superconductor — and also to develop a low-temperature specimen stage for an electron microscope. The experiments[21] were carried out as follows.

A toroidal magnet made of permalloy was covered with a superconducting niobium layer (see Fig. 3) to prevent electrons from penetrating into the magnet and also to confine the magnetic flux within the superconductor by exploiting the Meissner effect. Therefore, no magnetic field leaked out of the sample. The tiny samples were fabricated with the help of researchers working on Josephson-devices at our laboratory. As Yang predicted, we were able to check — by detecting the flux quantization phenomena — whether the covering superconductor actually became superconducting and whether it covered the magnet completely.

We fabricated many toroids of different sizes with various magnetic flux values (see Fig. 4) and measured the relative phase shifts. Nevertheless, as shown in Fig. 5, we obtained only two values for the relative phase shift: 0 and π.[21] A quantized flux with the value of $h/2e$ produces an electron phase shift of π. Therefore, when

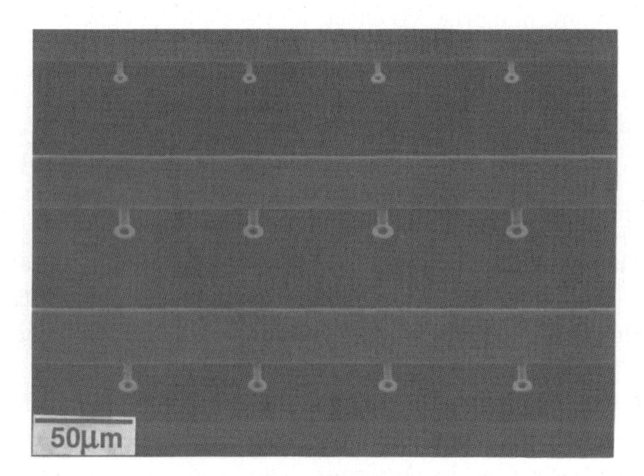

Fig. 4. Scanning electron micrograph of many toroidal samples.

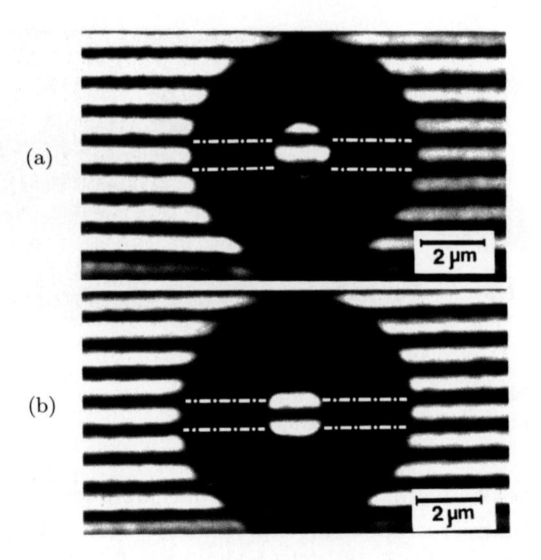

Fig. 5. Electron interference patterns showing the existence of the AB effect. (a) phase shift = 0; (b) phase shift = π.

an even number of flux quanta are trapped in the toroid, the phase shift becomes 0. When an odd number of flux quanta are trapped there, it becomes π.

The latter case confirmed the AB effect under the condition where an electron wave did not overlap with magnetic fields. Although an electron wave passed only through field-free regions, an observable effect was detected as a displacement of interference fringes. Vector potentials must have produced such an effect. The quantization of phase assure the complete covering with the superconductor and, consequently, magnetic shielding. Because of the Meissner effect, no magnetic fields leaked out.

In addition to confirming the AB effect, we were able to demonstrate the process of magnetic flux being quantized when the temperature crossed the critical temperature. When the temperature of the sample was 15 K, which is just above the critical temperature T_c, the niobium layer was normal. The fringe position inside the hole in Fig. 6(a) is a little bit lower than that of the middle line between two outside fringes. The relative phase shift determined by the magnetic flux flowing inside the toroid was 0.8π.

The phase shift suddenly increased to exactly π when the temperature decreased and crossed T_c [Fig. 6(b)]. When the niobium became superconductive, the supercurrent was induced to circulate around the magnet so that the total magnetic flux was quantized.

This result was published in Physical Review Letters[21] just before the second ISQM-Tokyo held in 1986. This symposium continues to be held every three years, and C. N. Yang has been on its Advisory Board and a key speaker from the first meeting through the fifth.[11,20]

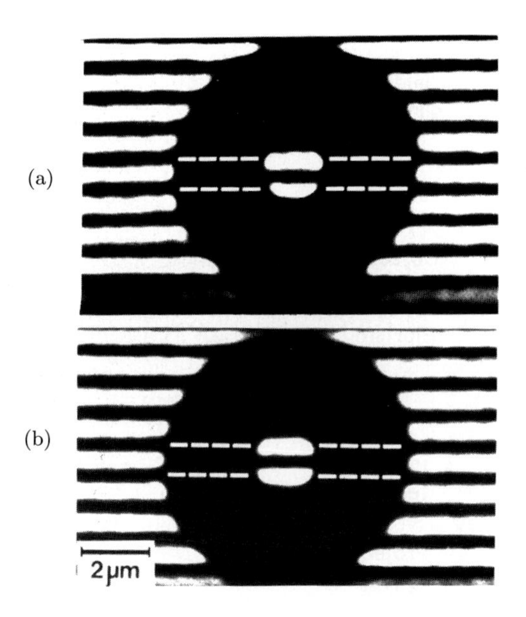

Fig. 6. Temperature dependence of interferograms. (a) Phase shift $= 0.8\pi$ at 15 K; (b) phase shift $= \pi$ at 5 K. The relative phase shift between a beam passing through the hole and a beam passing outside the toroid is determined by the magnetic flux flowing inside the toroid. When the sample temperature decreases below $T_c (= 9.2$ K), the phase shift suddenly changes to exactly π.

The principle behind the AB effect was used to practical applications such as the observation of the microscopic distribution of magnetic lines of force as will be explained below.

4. Using Holographic Electron Interference Microscopy to Observe Magnetic Lines of Force

A bright field-emission electron beam has facilitated practical applications of electron holography.[6] Electron holography consists of two steps: hologram formation and image reconstruction. In the first step, a hologram, which is an interference pattern between an object wave and a reference wave, is formed with an electron beam. An experimental arrangement for forming an off-axis electron hologram is shown in Fig. 7. An electron beam from a field-emission gun is collimated through a condenser lens system to illuminate an object. The collimated illumination is necessary if we are to obtain an illumination angle small enough for the spatial coherence length to cover both the object field and the reference beam. The image of the object is formed through an objective lens. When a positive potential is applied to the central filament of the electron biprism, the object image and the reference beam overlap each other to form an interference pattern. Since the interference fringe spacing is only a few hundred angstroms, the pattern is enlarged by magnifying lenses before it is recorded on film.

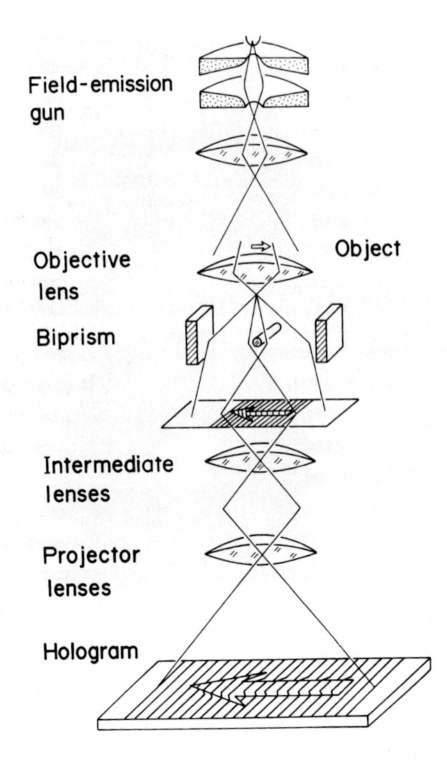

Fig. 7. Ray diagram for forming an off-axis electron hologram.

An optical image is reconstructed in one of the two diffracted beams simply by illuminating an off-axis electron hologram with a collimated laser beam corresponding to the reference beam used for hologram formation. This optical image is, except for lateral and transverse magnifications, essentially the same as the original electron image: the amplitude and phase are the same for both images. That is, the electron wavefronts are reproduced as optical wavefronts. In the other diffracted beam is produced a similar "conjugate" image whose amplitude is a complex-conjugate of the original amplitude.

Once the electron wavefronts are transformed into optical wavefronts, it is simple to optically read the phase distribution of an electron beam. In conventional interference microscopy, the precision of the phase measurement is approximately $2\pi/4$ and this is not always high enough to measure the electron phase shifts produced by microscopic objects. This is particularly true when an extremely small change in an object thickness or a small magnetic flux is to be observed.

The phase distribution is displayed as an interference micrograph by combining an optical interferometer with the optical reconstruction system. A conventional interference micrograph is formed by the interference between a reconstructed image and a plane wave, and the phase difference between the reconstructed wavefront and the plane wave produces a conventional interference micrograph as illustrated in

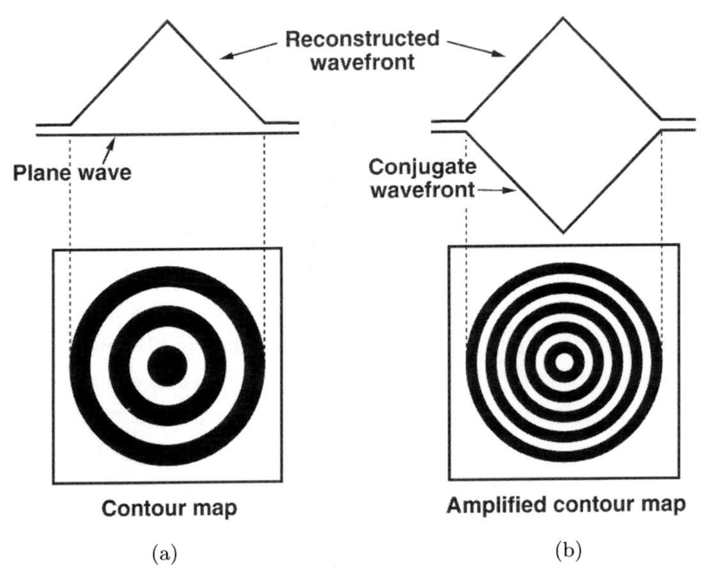

Fig. 8. Principle behind phase-amplified interference microscopy. (a) Contour map; (b) twice-amplified contour map.

Fig. 8(a). If a conjugate image instead of a plane wave overlaps the reconstructed wavefront, the phase difference becomes twice as large, as if the phase distribution were amplified two times. This is shown schematically in Fig. 8(b). The phase difference exactly doubles because twin images have amplitudes that are the complex conjugates of each other's: that is, their phase values are reversed in sign. As illustrated in Fig. 8(b), contour fringes thus appear at every phase change of π.

The problem is how to superimpose the twin images. An optical reconstruction system doing this is not difficult to design, and one way to obtain a phase-amplified image from an off-axis image hologram is shown in Fig. 9. A Mach–Zehnder interferometer splits a collimated light beam from a laser into two coherent beams (A and B) traveling in different directions. These two beams illuminate an electron

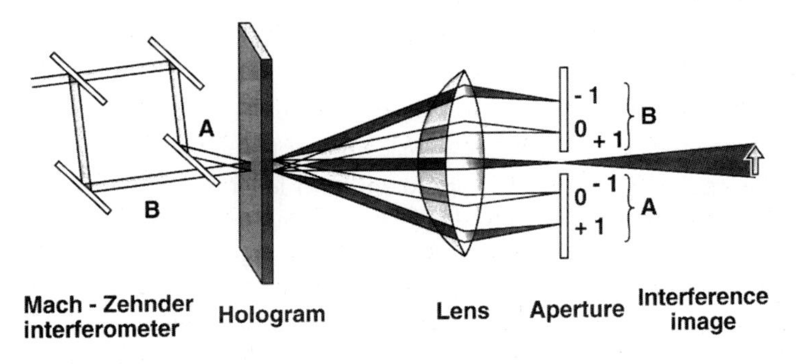

Fig. 9. Optical reconstruction system for phase-amplified interference microscopy.

hologram and produce two sets of a reconstructed image and its conjugate. The reconstructed image of one beam and the conjugate image of the other beam are made to coincide by using the Mach–Zehnder interferometer to adjust the incidence angles to the hologram. The phase distributions of these twin images are equal in absolute value but opposite in sign, resulting in an interference image with a two-fold phase distribution.

More than two-fold phase amplifications can be produced by repeating this process and also by using higher-order diffracted beams produced from a nonlinear hologram.

We found in 1980[7] that when a magnetic object is observed, its interference micrograph can be interpreted straightforwardly. When two electron beams starting from a point pass different points of a magnetic object and recombine at another point, the relative phase shift is given by Eq. (1). According to Stokes' theorem, this phase shift is determined by the magnetic flux Φ enclosed by the two paths:

$$\Delta\phi = -\frac{e}{\hbar}\oint \boldsymbol{A}\,ds = -\frac{e}{\hbar}\int \boldsymbol{B}\,d\boldsymbol{S} = -\frac{e}{\hbar}\Phi\,. \tag{2}$$

From this equation, we reach simple and clear conclusions:

(1) Contour fringes in an interference micrograph indicate projected magnetic lines of force, since no phase difference between two beams crossing a single magnetic line of force and therefore enclosing no magnetic flux is produced.
(2) These fringes are quantitative in that a constant minute magnetic flux, $h/e = 4 \times 10^{-15}$ Wb, flows between adjacent fringes.

An example of a ferromagnetic fine particle[7] is shown in Fig. 10. Although only the particle outline can be seen in the electron micrograph, two kinds of interference fringes appear in the interference micrograph. Contour fringes parallel to the edge indicate that thickness increases up to 550 Å from the edges; that is, the particle is a triangular pyramid truncated parallel to the base plane. The contour fringes in the central area enclosed by these sets of parallel fringes indicate magnetic lines of force, since thickness is uniform there. These fringes clearly show how the magnetization rotates inside the particle. They also provide quantitative information: since the phase is amplified twice in this micrograph, a magnetic flux of $h/2e$ flows between two adjacent contour fringes.

As illustrated in Fig. 10(a), this particle can be considered to consist of three magnetic domains. The observed magnetic lines of force, however, are almost circular since the radius of the particle is too small to form clearly separated domains.

Whether the magnetization is clockwise or counterclockwise cannot be determined from only this contour map. It can be determined from the interferogram, which is obtained when in the optical reconstruction stage a reference wave is tilted with respect to the object wave. Since the phase shift in this case was retarded at the central region of the particle, the rotation direction of the magnetization was found to be counterclockwise.

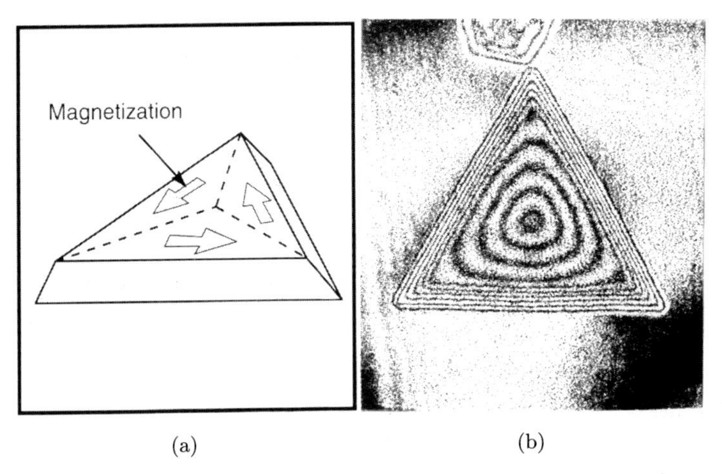

(a) (b)

Fig. 10. Cobalt fine particle. (a) Schematic of particle; (b) interference micrograph (phase amplification: $\times 2$).

More practical application fields are in magnetic recording, which is important in the storage of information for a variety of applications ranging from tape recorders to computer memories. To store larger amounts of information, we need to increase the recording density. And to do this we need microscopic observations of recorded patterns. An example of recorded patterns is shown in Fig. 11. In this in-plane magnetic recording, projected magnetic lines of force inside and outside the recorded cobalt film are observed.[22] Opposite magnetic lines in adjacent domains collide head-on with each other and produce vortex-like streams at the transition region. Since magnetic field $\boldsymbol{B}$ has neither a sink nor a source (div $\boldsymbol{B} = 0$), these magnetic

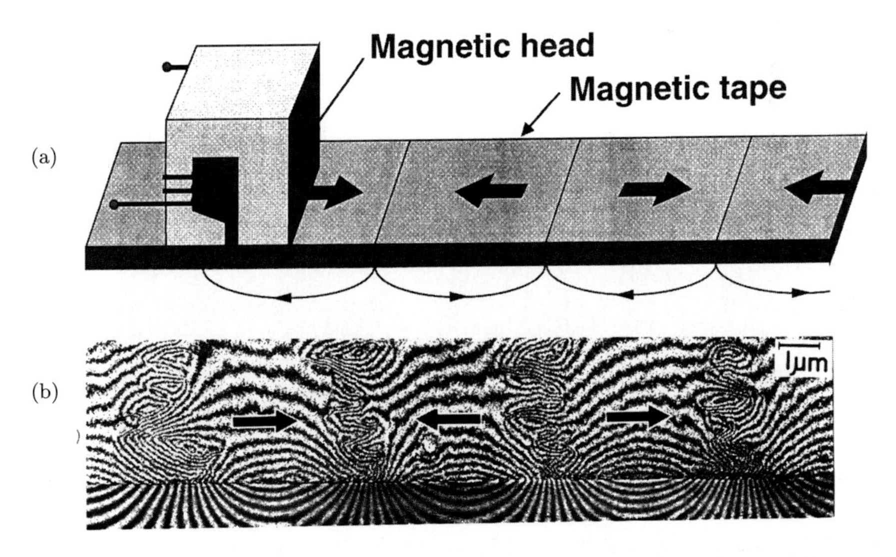

Fig. 11. Recorded magnetization patterns. (a) Recording method; (b) interference micrograph.

lines meander to the film edge and leak out. The narrowest width of the transition region, giving the limit of high-density recording, was confirmed with this method to be as small as 0.1 μm.[22]

Is it possible to observe magnetic lines of force inside a sample even when the sample is not uniform in thickness? The image of magnetic lines of force cannot be obtained from a single hologram. The phase distribution observed is the sum of the two contributions from the change in the sample thickness and the magnetic field. In case of the thickness change, the phase shift is produced by the electric effect. That is, when electrons penetrate a sample, they are accelerated by an amount determined by the inner potential of the sample. It was found[23] that we can separate the two contributions when we use different behavior of electrons by the time reversal operation. That is, Eq. (1) can be generalized to the case including the electric effect:

$$\Delta\phi = \left(\frac{1}{\hbar}\right) \oint (m\boldsymbol{v} - e\boldsymbol{A})ds\,. \tag{3}$$

If an electron beam is incident from the opposite direction ($t \to -t, \boldsymbol{v} \to -\boldsymbol{v}$, $s \to -s$), then the electric contribution to the phase shift [the first term in Eq. (3)] is the same but the magnetic contribution (the second term) reverses the sign of the phase. Two holograms were formed in the experiments discussed here: one in the standard specimen position, and one with the face of the specimen turned down. The electric contributions to the electron phase are the same in these holograms, but the magnetic contributions have opposite signs. We can easily understand this when we actually observe a three-dimensional ferromagnetic particle from two opposite directions.

Figure 12 shows two views of the interference micrographs of a cobalt decahedron. It is interesting to note that the phase images of the same sample differ between the top and bottom views, and this indicates the presence of extensive magnetic influence.

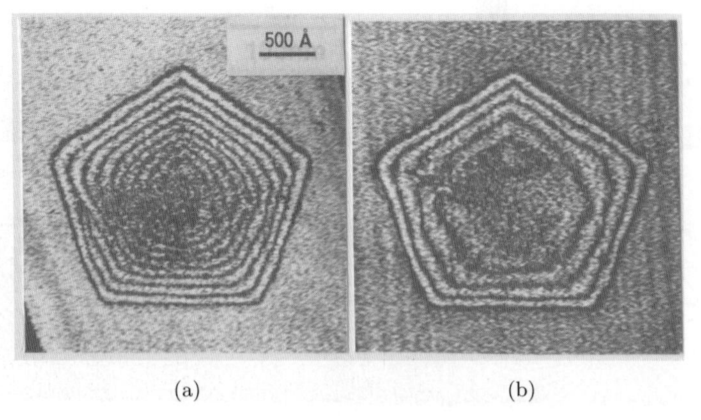

(a) (b)

Fig. 12. Two interference micrographs of a decahedral cobalt particle observed from opposite directions (phase amplification: ×2). (a) Top view; (b) bottom view.

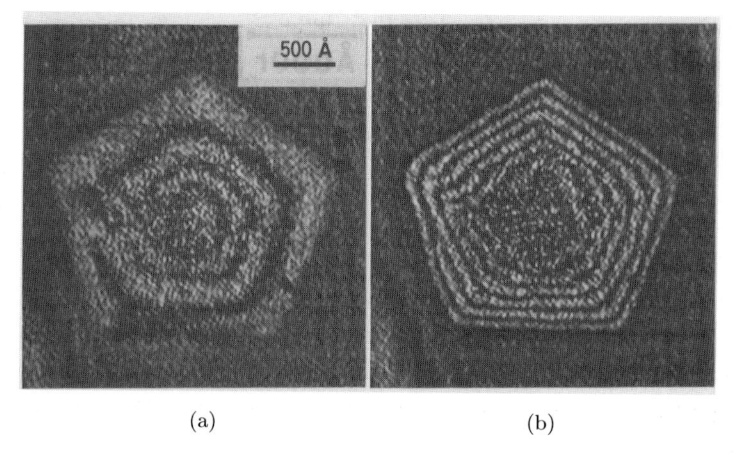

Fig. 13. Interference micrographs of a cobalt particle (phase amplification: ×2). (a) Magnetic lines of force (phase amplification: ×2); (b) thickness contour map (phase amplification: ×2).

If we consider the phase distribution in the micrograph shown in Fig. 12(a) to consist of an electric phase ($\Delta\phi_e$) plus a magnetic phase ($\Delta\phi_m$), then the phase distribution in the micrograph shown in Fig. 12(b) can be represented by the difference between the electric phase and the magnetic phase: $\Delta\phi_e - \Delta\phi_m$. If these two reconstructed images are made to overlap in order to show the phase difference between them, the resultant phase distribution becomes two times the magnetic phase: $2\Delta\phi_m$.

In this way, a purely magnetic image is obtained. A thickness image, on the other hand, is obtained when we make full use of the conjugate image, which has a phase distribution with a sign opposite the original sign. The pure thickness image ($2\Delta\phi_e$) can be obtained by overlapping the reconstructed image of the top view and the conjugate image of the bottom view.

The results are shown in Fig. 13, where micrographs (a) and (b) show magnetic lines of force and thickness contours, respectively. Although many choices are available for the domain structure of a three-dimensional particle, the magnetization in this sample can be determined to rotate around the side common to five regular tetrahedrons forming a particle like that shown in Fig. 14.

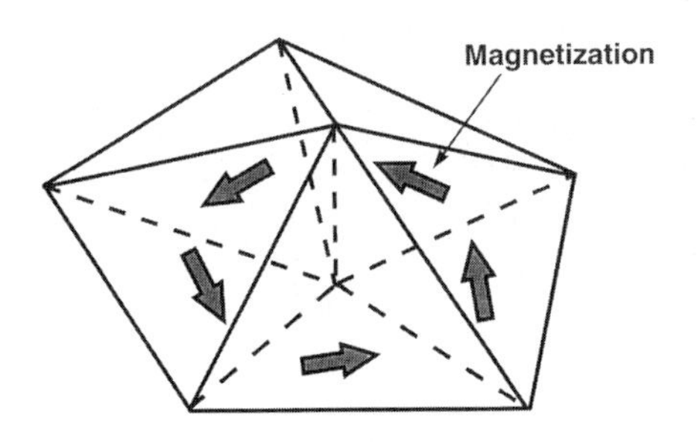

Fig. 14. Schematic diagram of the magnetic domain structure in a cobalt particle.

The techniques developed for detecting slight phase shifts by using holographic techniques and for observing the AB effect by using a toroidal magnet covered with a superconductor were used to observe quantized fluxes, or vortices, penetrating superconductors.

5. Real-Time Observation of Individual Vortices in Superconductors

Quantized vortices in a superconductor play an important role in both the fundamentals and practical applications of superconductivity. When a magnetic field is applied to a type-II superconductor, quantized vortices are produced in the form of extremely thin filaments of magnetic flux penetrating it. The magnetic flux of a vortex is quantized because the phase of the Cooper-pair wave circulating around the filament changes by 2π after a full circuit. A vortex should have a normal core in its center to avoid phase singularity just like the eye of a hurricane. These tiny vortices hold the key when superconductors are to be practically used for dissipation-free conductors: when a current is applied, vortices receive a Lorentz force and begin to move. Since the movement of magnetic flux induces a voltage difference, the superconducting state is destroyed. To obtain a large dissipation-free current, we therefore have to fix or pin down the vortices. Vortices were found experimentally to be pinned at material defects where superconductivity is locally weak or broken, since the vortex core is more stable at such defects than at other places. The microscopic mechanism of vortex pinning, however, is not fully known.

Individual vortices were first imaged by using the Bitter technique[24] and have since been imaged by using other, newly developed techniques.[25–27] We have not, however, been able to observe in real time when and how vortices are pinned at defects and how they are depinned from them when a driving force is applied. We tried to observe the dynamics of vortices directly and dynamically to derive the guiding principle for developing high critical-current superconductors. In 1989 we first observed magnetic lines of force of vortices leaking from the superconductor surface.[28]

5.1. *Magnetic lines of force of vortices leaking from the surface of a superconductor*

Magnetic lines of force of vortices form thin filaments inside a superconductor and become uniform outside it. However, magnetic lines of force near the superconductor surface are still localized even outside the superconductor. We observed these magnetic lines of force by electron interference microscopy using the experimental arrangement illustrated in Fig. 15. A thin film of lead was prepared by evaporating lead onto one side of a wire, and a magnetic field of a few gauss was applied perpendicular to the film. When the specimen was then cooled to 4.5 K, the magnetic lines penetrated the superconductor in the form of vortices. An electron beam was

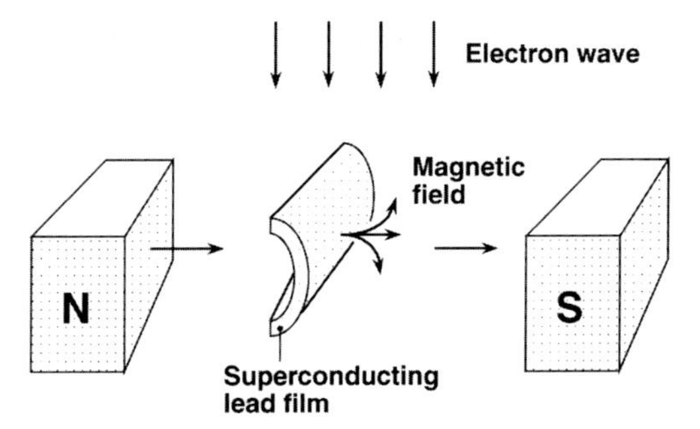

Fig. 15. Experimental arrangement for vortex observation. A magnetic field is applied perpendicular to a thin film of lead. Magnetic lines of force penetrating the film are observed through electron holography with an electron beam incident from above.

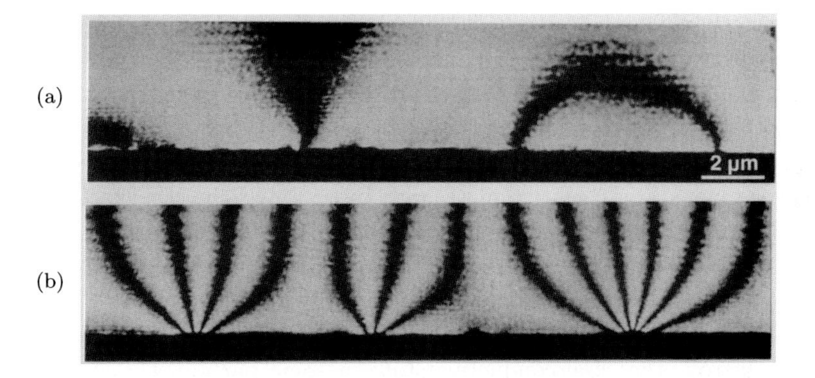

Fig. 16. Interference micrographs of magnetic lines of force penetrating thin films of lead (phase amplification: $\times 2$). One fringe corresponds to a flux of $h/2e$ (i.e. to the flux of a single vortex). (a) Film thickness = 0.2 μm. A magnetic line of force is produced from a single vortex in the sample surface on the right-hand side of the micrograph. An antiparallel pair of vortices is seen on the left. (b) Film thickness = 1.0 μm. A bundle of vortices is penetrating a lead film ("intermediate state").

incident onto the specimen from above and the magnetic lines sticking out from the superconductor surface were observed.

Figure 16 shows the resultant interference micrographs for superconductive lead films 0.2 μm and 1.0 μm thick.[28] Since the micrographs are phase-amplified by a factor of 2, one contour fringe corresponds to one vortex. A magnetic line from a single vortex can be seen in the right-hand side of the micrograph shown in Fig. 16(a). The magnetic line is produced from an extremely small area of the lead surface, and it then spreads out into the free space.

If the distribution of the magnetic field of the vortex is assumed to be axially symmetric, the magnetic field near the surface of the specimen can be estimated from the micrograph to be as high as 1000 G even though the applied magnetic field is only a few gauss. This is because a tiny and strong solenoid is formed by the superconducting vortex current.

Although these fringes could be attributed to vortices since each fringe corresponds to a magnetic flux of $h/2e$, further confirmation was obtained as follows: when the specimen temperature was raised above the critical temperature T_c, these magnetic lines disappeared. This is interpreted as a sign that the superconducting state was destroyed above T_c, causing the vortex current to stop flowing. This is evidence that the magnetic lines observed were generated by the superconductive vortex current.

We also observed a pair of vortices oriented in opposite directions and connected by the magnetic lines of force [Fig. 16(a), left]. One possible reason for the pair's creation is that when the specimen was cooled below the critical temperature the lead was brought into the superconductive state. During the cooling process, however, the specimen was in a state in which the vortex pair appeared and disappeared repeatedly because of thermal excitations. This phenomenon, which was predicted by the Kosterlitz–Thouless theory,[29,30] is peculiar to two-dimensional systems such as thin films or layered structures and is expected to occur in layered superconductors such as high-T_c superconductors. During the cooling process, an antiparallel pair of vortices is produced, pinned by some imperfections in the superconductor, and eventually frozen.

What happens in a thicker film? Figure 16(b) shows the state of the magnetic lines of force observed when the film was 1 μm thick. The state is completely changed: magnetic flux penetrates the superconductor not individually in vortex form, but in a bundle. No vortex pairs are seen in this figure.

Since lead is a type-I superconductor, a strong magnetic field applied to it partially destroys the superconductive state in some parts of the specimen ("intermediate state"). The magnetic lines of force seen in Fig. 16(b) penetrate those parts of the specimen where superconductivity is destroyed, but because the surrounding regions are superconductive the total amount of penetrating magnetic flux is quantized to an integral multiple of the flux quantum $h/2e$. A superconducting lead film less than 0.5 μm thick, like that shown in Fig. 16(a), behaves like a type-II superconductor and the flux penetrates the superconductor in the form of individual vortices.

5.2. *Dynamic observation of vortices at defects*

To directly observe the vortex pinning phenomena microscopically, we need to observe in real time how individual vortices behave at material defects. The arrangement[31] for this experiment is shown in Fig. 17: a superconducting thin film is tilted to the incident electrons, and an external magnetic field is applied. The magnetic fields of vortices deflect, or phase-shift, the electrons passing through the film. Therefore, we can see a vortex as a pair of black-and-white spots in the lower out-of-focus plane (Lorentz micrograph). We can thus observe individual vortices interacting with defects in real time, although the defect images are blurred because of the image defocusing.

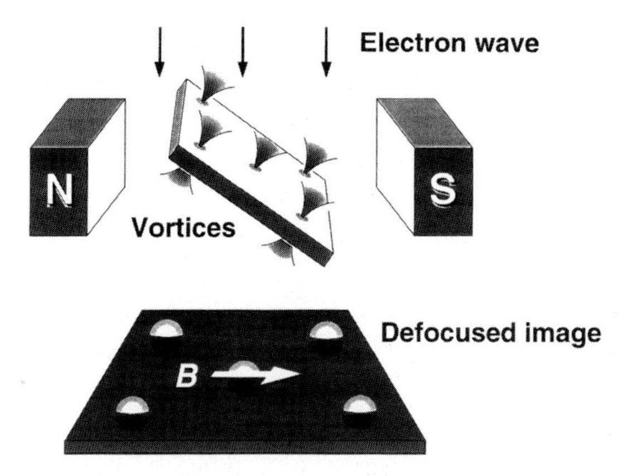

Fig. 17. Principle behind Lorentz microscopy of vortices. An incident electron wave is phase-shifted by the magnetic fields of vortices. When the image is greatly defocused, the phase shifts are manifested as intensity variations.

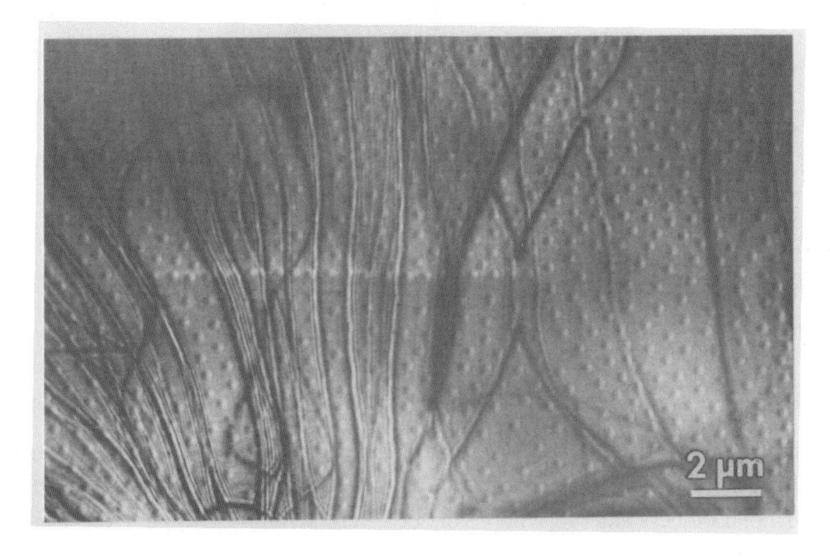

Fig. 18. Lorentz micrograph showing the interaction of vortices with a surface step in Nb thin film ($T = 4.5$ K and $B = 20$ G).

When the magnetic field applied to a niobium thin film that had surface steps and was at 4.5 K changed, vortices began to move until they reached the equilibrium state. Vortices arriving at a surface step did not get over the step. However, when there was a nonzero component of the vortex motion, vortices turned to move along the step. A Lorentz micrograph obtained when the vortex movement was completed is shown in Fig. 18. At first glance, vortices appear to be distributed at random, but when you carefully see, you may notice that vortices bunch up along the surface step running in the perpendicular direction in the photograph. It is

Fig. 19. Vortices in a Nb thin film having a regular array of artificial defects. Vortices are strongly pinned at artificial defects (black dots). Consequently vortices do not form a regular lattice but instead form domains of lattices.

evident that a surface step acts as a pinning line when the direction of the vortex motion is perpendicular to it and as a channel when the direction of motion is along the step.

5.3. *Plastic vortex-movements at artificial defects*

The dynamic interaction of vortices with point defects in niobium thin films was also investigated. Artificial defects were produced by focused ion-beam irradiation. When artificial defects existed where vortices were strongly trapped, vortices could not form a regular lattice and divided into domains of lattices. As can be seen in Fig. 19, the domain boundaries and dislocations in vortex lattice were produced near the "black defects." When a force was applied by changing the applied magnetic field, the vortices resisted moving for a while until some vortices began to flow in "avalanches," resulting in "vortex rivers" along the domain boundaries.[32] When vortices flowed, they became disordered. The flow was not continuous but intermittent. When they stopped flowing, however, the vortices formed a new configuration of lattices, and then "vortex rivers" began to flow along the new domain boundaries. This process was repeated until vortices reached the equilibrium configuration.

The vortex movement was very different when a regular and dense array of defects was produced in the film: at specific values of magnetic fields, vortices formed regular and rigid configurations determined by the lattice of defects.[33] Under

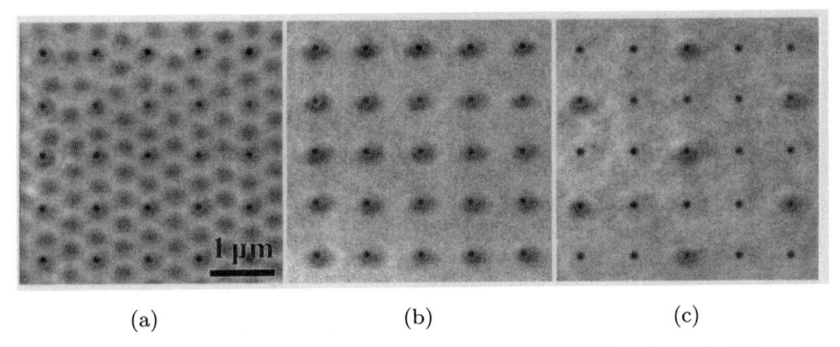

Fig. 20. Vortices in a Nb thin film having a dense square array of artificial defects (H_1: matching magnetic field). (a) $H = 4H_1$; (b) $H = H_1$; (c) $H = (1/4)H_1$.

this condition, vortices did not start to move easily. This vortex pinning behavior corresponds to the peak effect or the matching effect indicating that the critical current increases at specific values of the applied magnetic field, which was observed using macroscopic measurements.

Some Lorentz micrographs showing the configurations of vortices relative to defect positions are shown in Fig. 20. The locations of the defects are indicated by black dots in the micrographs. At the matching magnetic field, H_1 [Fig. 20(b)], all the defects were occupied by vortices without any vacancies or interstices, thus forming a rigid square lattice. The peak effect of the critical current can be explained microscopically: when vortices form a stable and regular lattice, even if a vortex is depinned from one pinning site it cannot find any vacant site to move to. As a result, these vortices require more force to start moving.

Regular lattices were formed not only at $H = H_1$, but also at $H = mH_1/n$ (n, m; integer). For example, as in the case of $H = 4H_1$ [see Fig. 20(a)], defect positions forming a square lattice were first occupied by vortices. Then two vortices aligned in the vertical direction were inserted at every interstitial site, and finally an additional vortex was inserted midway between two adjacent defects in the vertical direction. Figure 20(c) shows the case at $H = 1/4H_1$. Vortices occupy every fourth site in the horizontal direction thus forming a centered (4×2) rectangle lattice. The reason the pinning force as a whole becomes stronger at the specific values of magnetic fields is that vortices form rigid and regular lattices.

When "excess" (or "deficient") vortices were produced at magnetic fields different from the specific ones, such excess vortices (or vacant vortices) did not form a regular lattice and therefore hopped by a weaker force just like the conduction of "electrons" and "holes" in a semiconductor.

5.4. *Onset of vortex movements in high-T_c superconductors*

Vortices in high-temperature superconductors were difficult to observe, because the magnetic radius of vortices, which is equal to penetration depth λ, is much larger in high-temperature superconductors than in Nb. In addition, the radius increases further since the film thickness observable by using 300-kV electrons is smaller than λ. However, we made several efforts to dynamically observe vortices in high-T_c superconductors:[34] the image contrast was enhanced by increasing a film thickness to shorten the effective λ, and by increasing the tilting angle α of the film (see Fig. 17) to obtain a larger phase shift of electrons due to a vortex.[35] The image brightness was also increased by using a beam intensity an order of magnitude greater than previously used to observe vortices in Nb.

Thin-film samples of Bi-2212 ($T_c = 85$ K) were prepared by peeling them off a single crystal. A driving force was applied to vortices by changing the magnetic field H, and the onset of the vortex movement was observed. When we observed how the vortices moved when $0 \text{ G} \leqq H \leqq 45$ G and $7 \text{ K} \leqq T \leqq 50$ K, we found that their manner of movement changed markedly depending on H and T. Among others, the dynamic behavior of vortices was completely different above and below 25 K: at $T < 25$ K all the vortices migrated slowly and at almost the same speed. The video frame shown in Fig. 21(a) was obtained $1s$ before that shown in Fig. 21(b). This pair of Lorentz micrographs thus indicates that at $T = 20$ K and $dH/dt = 0.008$ G/s the vortex speed was 1.5 μm/s. The speed decreased rapidly as T decreased. The speed increased as dH/dt increased.

We interpreted that this migration movement is due to the fact that in high-Tc superconductors, vortices can be pinned even at extremely small, densely-

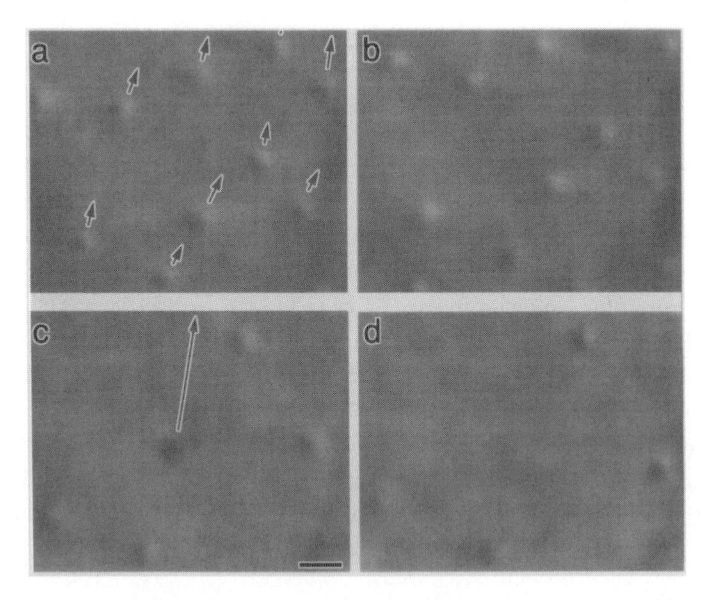

Fig. 21. Lorentz micrographs showing the movement of vortices in a high-T_c Bi-2212 film. (a)–(b) Migration movement at 20 K. (c)–(d) Hopping movement at 30 K. The vortex movement when the force F is exerted on vortices by changing H (at $dH/dt = 0.008$ G/s) differs between at low and high temperatures. Below 25 K, all the vortices migrate almost in parallel. Above 25 K, vortices suddenly hop from one pinning center to another. The Lorentz micrograph in (a) was obtained 1 s before that shown in (b).

distributed oxygen defects due to the small core radius of a vortex: since a single vortex penetrating a 2000 Å-thick film may be collectively pinned by more than 100 oxygen defects, vortices would apparently move smoothly just like in a viscous medium when thermally activated.

At $T > 25$ K, pinning centers began to appear and vortices were no longer able to migrate [Fig. 21(c)]. The vortices instead began to be trapped at the pinning centers and started to suddenly hop when the driving force increased [Fig. 8(d)], and the vacated pinning centers were soon occupied by newly arriving vortices. The hopping speed was so high that the vortex image on the video screen looked as if it were blinking on and off.

The reason the behavior of vortex motion changed at 25 K can be explained as follows: the collective pinning due to tiny oxygen defects decreased rapidly with increasing T because the increased thermal vibration of vortices easily depinned them from tiny atomic pinning centers. The pinning of vortices at larger and sparser defects, which have not yet been identified, became dominant instead.

Even when the vortices became more closely spaced ($H > 1$ G) and the vortex–vortex interaction became stronger, vortices kept migrating until $H = 45$ G. At $T > 25$ K, however, various forms of vortex motion — such as filamentary flow, river flow and lattice-domain flow — were observed when we increased H.

6. Conclusions

The development of a bright field-emission electron beam made it possible not only to investigate the wave nature of electrons but also to observe the hitherto unobservable microscopic phenomena. The Aharonov–Bohm effect, the most fundamental principle underlying the interaction of electron waves with electromagnetism, was confirmed conclusively through a series of experiments using this electron beam and photolithography techniques. Furthermore, this AB effect principle was used to practical applications to the observation of the microscopic distributions of electromagnetic fields. Magnetic lines of force were directly displayed as phase contour fringes in electron interference micrographs obtained using the electron holography process, and quantized vortices penetrating a thin film of type-II superconductor were dynamically observed as spots of black-and-white contrast features in Lorentz micrographs.

For these past 30 years, new ways of utilizing the phase information of electron waves have become possible every time a brighter electron beam was developed. The coherence of an electron beam, however, cannot still be compared to that of a laser. Therefore, we developed a 1000-kV field-emission electron microscope[36] which had a higher-energy and brighter electron beam.

We think that we need not to stress that brighter sources bring about new possibilities in a wide range of fields in science and technology, as examples of lasers, synchrotron radiation and neutron sources did. This new microscope, will, we hope, help expand the field of electron phase microscopy, thus making feasible new fundamental experiments in quantum mechanics as well as new applications for ultrafine measurements.

References

1. A. Tonomura, T. Matsuda, J. Endo, H. Todokoro and T. Komoda, *J. Electron Microsc.* **28**, 1 (1979).
2. A. V. Crewe, D. N. Eggenberger, D. N. Wall and L. M. Welter, *Rev. Sci. Instrum.* **39**, 576 (1968).
3. A. V. Crewe, J. Wall and J. Langmore, *Science* **168**, 1338 (1970).
4. D. Gabor, *Proc. R. Soc. London* **A197**, 454 (1949).
5. A. Tonomura, A. Fukuhara, H. Watanabe and T. Komoda, *Jpn. J. Appl. Phys.* **7**, 295 (1968).
6. A. Tonomura, *Electron Holography*, 2nd ed. (Springer, Heidelberg, 1999).
7. A. Tonomura, T. Matsuda, J. Endo, T. Arii and K. Mihama, *Phys. Rev. Lett.* **44**, 1430 (1980).
8. M. Peshkin and A. Tonomura, *Lecture Notes in Physics* (Springer-Verlag, 1989), Vol. 340.
9. Y. Aharonov and D. Bohm, *Phys. Rev.* **115**, 485 (1959).
10. T. T. Wu and C. N. Yang, *Phys. Rev.* **115**, 485 (1975).
11. For example, C. N. Yang, *Proc. 5th Int. Symp. on Foundations of Quantum Mechanics in the Light of New Technology (ISQM-Tokyo '95)* (Elsevier, Amsterdam, 1996), p. 307.
12. C. N. Yang and R. L. Mills, *Phys. Rev.* **96**, 191 (1954).
13. R. G. Chambers, *Phys. Rev. Lett.* **5**, 3 (1960).
14. P. Bocchieri and A. Loinger, *Nuovo Cimento* **A47**, 475 (1978).
15. For example, C. G. Kuper, *Phys. Lett.* **794**, 413 (1980).
16. See F. Dyson, *Mod. Phys. Lett.* **A14**, 1 (1999).
17. C. N. Yang, *Selected Papers 1945–1980 With Commentary* (W. H. Freeman and Company, New York, 1983), p. 305.
18. A. Tonomura, T. Matsuda, R. Suzuki, A. Fukuhara, N. Osakabe, H. Umezaki, J. Endo, K. Shinagawa, Y. Sugita and H. Fujiwara, *Phys. Rev. Lett.* **48**, 1443 (1982).
19. P. Bocchieri, A. Loinger and G. Siragusa, *Lett. Nuovo Cimento* **35**, 370 (1982).
20. See *Foundations of Quantum Mechanics in the Light of New Technology*, eds. S. Nakajima, Y. Murayama and A. Tonomura, *Advanced Series in Applied Physics* (World Scientific, Singapore, 1996), Vol. 4, p. 25.
21. A. Tonomura, N. Osakabe, T. Matsuda, T. Kawasaki, J. Endo, S. Yano and H. Yamada, *Phys. Rev. Lett.* **56**, 792 (1986).
22. N. Osakabe, K. Yoshida, Y. Horiuchi, T. Matsuda, H. Tanabe, T. Okuwaki, J. Endo, H. Fujiwara and A. Tonomura, *Appl. Phys. Lett.* **42**, 746 (1983).
23. A. Tonomura, T. Matsuda, J. Endo, T. Arii and K. Mihama, *Phys. Rev.* **B34**, 3397 (1986).
24. U. Essman and H. Träuble, *Phys. Lett.* **24**, 526 (1967).
25. H. F. Hess, R. B. Robinson and J. V. Waszczak, *Phys. Rev. Lett.* **64**, 2711 (1990).
26. A. M. Chang, H. D. Hallen, L. Harriott, H. F. Hess, H. L. Kao, J. Kwo, R. E. Miller, R. Wolfe, J. van der Ziel and T. Y. Chang, *Appl. Phys. Lett.* **61**, 1974 (1992).
27. L. N. Vu, M. S. Wistrom, D. J. Van Harlingen, *Appl. Phys. Lett.* **63**, 1693 (1993).
28. T. Matsuda, S. Hasegawa, M. Igarashi, T. Kobayashi, M. Naito, H. Kajiyama, J. Endo, N. Osakabe, A. Tonomura and R. Aoki, *Phys. Rev. Lett.* **62**, 2519 (1989).

29. J. M. Kosterlitz and D. J. Thouless, *J. Phys.* **6**, 1181 (1973).
30. B. I. Halperin and R. Nelson, *J. Low Temp. Phys.* **36**, 599 (1979).
31. K. Harada, T. Matsuda, J. E. Bonevich, M. Igarashi, S. Kondo, G. Pozzi, U. Kawabe and A. Tonomura, *Nature* **360**, 51 (1992).
32. T. Matsuda, K. Harada, H. Kasai, O. Kamimura and A. Tonomura, *Science* **271**, 1393 (1996).
33. K. Harada, O. Kamimura, H. Kasai, T. Matsuda and A. Tonomura, *Science* **274**, 1167 (1996).
34. A. Tonomura, H. Kasai, O. Kamimura, T. Matsuda, K. Harada, J. Shimoyama, K. Kishio and K. Kitazawa, *Nature* **397**, 308 (1999).
35. J. E. Bonevich, K. Harada, T. Matsuda, H. Kasai, T. Yoshida, G. Pozzi and A. Tonomura, *J. Electron Microsc.* **42**, 88 (1993).
36. T. Kawasaki, T. Yoshida, T. Matsuda, N. Osakabe, A. Tonomura, I. Matsui and K. Kitazawa, *Appl. Phys. Lett.* **76**, 1342 (2000).

Ernest D. Courant with C. N. Yang and M. Goldhaber

POSSIBILITIES FOR SPIN PHYSICS AT HIGH ENERGY

ERNEST D. COURANT

Brookhaven National Laboratory, Upton, NY 11973, USA

1. Introduction

Since elementary particles have spin, it behooves physicists to explore whether and to what extent their behavior and interactions depend on that variable. Accordingly it is of interest to do experiments with polarized particle beams.

To produce a beam of high energy polarized particles, one may conceive of two approaches:

(a) Accelerate an unpolarized beam, then manipulate it so as to polarize it.
(b) Polarize the particles in an ion source, then accelerate them.

For electrons, approach (a) is the most practical. This is due to the Sokolov–Ternov mechanism, found by two Soviet physicists in the 1960's.

Electrons circulating in a magnetic field emit synchrotron radiation. When a quantum of synchrotron light is emitted, the electron's spin usually stays the same, but for a small fraction of the time it reverses. It turns out that the rate at which electrons radiate with the spin flipping from parallel to the magnetic field to antiparallel is greater than the rate in the opposite direction:

$$W_+ = W_{\downarrow\uparrow} = \frac{1}{2} W_0 \left(1 - \frac{8}{5\sqrt{3}}\right) \tag{1}$$

$$W_- = W_{\uparrow\downarrow} = \frac{1}{2} W_0 \left(1 + \frac{8}{5\sqrt{3}}\right) \tag{2}$$

$$P = \frac{W_+ - W_-}{W_+ + W_-} = -\frac{8}{5\sqrt{3}} = -92.4\% \tag{3}$$

where W_0 is the average spin flip rate.

The time constant τ for approach to this equilibrium polarization is

$$\tau = \frac{8}{5\sqrt{3} \times 137\gamma^5} \left(\frac{\rho}{r_0}\right)^2 \frac{R}{c} \tag{4}$$

where ρ is the radius of orbit curvature, r_0 is the classical electron radius, and R is the average machine radius. As an example, for LEP (100 GeV) with $\gamma = 2 \times 10^5$, $\rho = 3.1$ km, $R = 4.24$ km we find

$$\tau = 6 \text{ min}$$

which is quite reasonable. But with protons the situation is different: in RHIC, we have $\gamma = 250$, $R = 600$ m, $\rho = 280$ m, and the classical radius is that of the proton. Now the polarization time appears to be 1.5×10^{13} years — longer than the life of the universe, but less than the lifetime of the proton. So this mechanism will not work for protons. We must create polarized protons in an ion source and then accelerate.

2. Proton Spin Dynamics

Ion sources are available; I have no time to discuss them. During the acceleration process — in the magnetic field of a circular accelerator — the spin will precess according to the famous Thomas–BMT equation

$$\frac{d\mathbf{S}}{dt} = \mathbf{S} \times \mathbf{\Omega} = \frac{e}{m\gamma}\mathbf{S} \times [(1+\gamma G)\mathbf{B}_{\rm tr} + (1+G)\mathbf{B}_{\rm long}] \tag{5}$$

where $G = 1.7928$ is the anomalous magnetic moment coefficient of the proton, and $B_{\rm tr}$ and $B_{\rm long}$ are the components of the magnetic field transverse and parallel to the particle velocity.

One way to look at this is to say that the anomalous moment effectively transforms proportional to the energy.

The effect is that the ratio $\nu_{\rm sp}$ of the spin precession frequency to the orbital revolution frequency increases with energy:

$$\nu_{\rm sp} = 1 + \gamma G\,. \tag{6}$$

In a uniform vertical magnetic field, therefore, the spin precesses about the vertical direction at the rate of (6) times (or, in a coordinate system rotating with the particle, γG times) the revolution frequency. In particular, the spin may be exactly vertical (aligned with the magnetic field) everywhere.

But the actual magnetic field is not exactly uniform and vertical; the field in (5) as seen by the particle also contains horizontal components. These arise from the focusing fields that govern vertical oscillations, as well as from possible alignment and construction errors. These fields will turn the spin away from the vertical. If these depolarizing field have a frequency component γG (in the reference frame of the particle) resonance arises, and the spin can be substantially changed, leading to depolarization.

3. Resonances

The horizontal fields due to misalignments and field errors produce a central orbit that deviates somewhat from the ideal obit; the deviations of the closed orbit and, with it, of the field contain all integral harmonics of the orbit frequency. This leads to depolarizing resonances whenever $\nu_s = k$ (k integral):

$$\gamma G = k \text{ (imperfection resonances)}\,. \tag{7}$$

The frequency of vertical oscillations is ν_s times the orbit frequency, where ν_s, called the *vertical tune*, is determined by the lattice (magnet configuration) of the ring. In an alternating-gradient accelerator or storage ring the oscillations are not purely sinusoidal but are modulated by the overall periodicity of the lattice (e.g. 12-fold for the Brookhaven AGS, 3-fold for RHIC, 6-fold for FNAL). Therefore the resonance condition for spin depolarization by vertical oscillations is

$$\gamma G = kP \pm \nu_s \text{ (intrinsic resonance)}. \tag{8}$$

(These resonances are called *intrinsic* because betatron oscillations inevitably take place in any ring.) Here k is any integer, P is the periodicity of the lattice structure.

To see how these resonances work, we reformulate the spin equation of motion in using the formulation of spinors and SU_2 matrices. The spin, in terms of the spinor σ, is

$$\mathbf{S} = \psi^{+}\boldsymbol{\sigma}\psi \tag{9}$$

and the equation of motion for ψ equivalent to (5) is

$$\frac{d\psi}{dt} = \frac{i}{2}\left(\boldsymbol{\sigma}\cdot\boldsymbol{\Omega}\right)\psi. \tag{10}$$

Changing the independent variable to the bending angle θ and moving to a coordinate system which rotates with the path of the particle, the equation becomes

$$\frac{d\psi}{d\theta} = \frac{i}{2}\begin{pmatrix} -\kappa & \varsigma(\theta) \\ \varsigma^*(\theta) & \kappa \end{pmatrix}\psi \tag{11}$$

with $\kappa = \gamma G$; ζ can be expressed in terms of the motion of the particle:

$$\varsigma = -(1+\gamma G)(\rho z'' + iz') + i\rho(1+G)\left(\frac{z}{\rho}\right)' + \text{higher order} \tag{12}$$

which can be expressed as a combination of oscillations:

$$\varsigma = \sum_k \varepsilon_k \exp(-i\kappa_k\theta). \tag{13}$$

Here ε_k is the *strength* of the resonance at frequency κ_k, where κ_k is either an integer or an intrinsic resonance frequency (8).

In the case of a perfect machine there is only a vertical field, and the solution of (11) is simply

$$\psi = \exp\left(-\frac{i}{2}\kappa\theta\sigma_3\right)\psi_0 \tag{14}$$

or, if κ varies with time,

$$\psi = \exp\left(-\frac{i}{2}\chi\sigma_3\right)\psi_0 = \begin{pmatrix} e^{-i\chi/2} & 0 \\ 0 & e^{i\chi/2} \end{pmatrix}\psi_0 \quad \text{with } \chi = \int \kappa(\theta)d\theta. \tag{15}$$

In this case the spin precesses about the vertical (z) direction at the rate κ, and the vertical component remain constant.

3.1. *Single resonance*

When the various frequencies κ_k are well separated, we may consider one of the components of (14) at a time, say with strength κ at frequency κ_0:

$$\frac{d\psi}{d\theta} = \frac{i}{2}\begin{pmatrix} -\kappa & \varepsilon e^{-i\kappa_0\theta} \\ \varepsilon e^{i\kappa_0\theta} & \kappa \end{pmatrix}\psi = \frac{i}{2}\left(-\kappa\sigma_3 + \varepsilon\sigma_1 e^{i\kappa_0\theta\sigma_3}\right)\psi\,. \tag{16}$$

Transform to a system rotating with the perturbation frequency κ_0:

$$\varphi = \exp\left(\frac{i}{2}\kappa_0\theta\sigma_3\right)\psi$$

and obtain

$$\frac{d\varphi}{d\theta} = \frac{i}{2}\begin{pmatrix} -\delta & \varepsilon \\ \varepsilon & \delta \end{pmatrix}\varphi = \frac{i}{2}(-\delta\sigma_3 + \varepsilon\sigma_1)\varphi \tag{17}$$

where $\delta = \kappa - \kappa_0$, the distance between the excitation frequency κ_0 and the resonance frequency κ.

Since this has constant coefficients, it is easily solved:

$$\varphi = \exp\left[\frac{i}{2}(-\delta\sigma_3 + \varepsilon\sigma_1)(\theta - \theta_0)\right]\varphi_0\,. \tag{18}$$

Transforming back to the spinor ψ

$$\psi = M\psi_0$$

with

$$M = \exp\left(-\frac{i}{2}\kappa_0\theta\sigma_3\right)\exp[i\pi(-\delta\sigma_3 + \varepsilon\sigma_1)(\theta - \theta_0)]\exp\left(\frac{i}{2}\kappa_0\theta_0\sigma_3\right)\,. \tag{19}$$

This spin transfer matrix M is a SU_2 matrix, and can be parameterized as

$$M = \exp(i\pi\lambda\mathbf{n}\cdot\boldsymbol{\sigma}) \text{ with } \lambda = \sqrt{\delta^2 + \varepsilon^2} \tag{20}$$

meaning that it represents a rotation by the angle $2\pi\lambda$ around a unit $\mathbf{n}$. If M is the matrix for one complete turn, $\mathbf{n}$ is the *invariant spin direction*; i.e. a spin vector $\mathbf{n}$ transforms into itself after one turn. In fact, *any* SU_2 matrix M can be parametrized in the form (20); the *spin tune* (number of rotations) is just

$$\nu_{\text{sp}} = \frac{1}{\pi}\arccos\left(\frac{1}{2}\operatorname{Tr} M\right)\,. \tag{21}$$

For a turn from θ to $\theta + 2\pi$ the vector $\mathbf{n}$ is

$$\mathbf{n} = \begin{pmatrix} (\varepsilon/\lambda)\cos\kappa_0(\pi + \theta) \\ (\varepsilon/\lambda)\sin\kappa_0(\pi + \theta) \\ \delta/\lambda \end{pmatrix} \tag{22}$$

i.e. the vertical component of the invariant spin is proportional to the distance from resonance, while the horizontal component precesses at the rate κ_0 . Therefore if δ is large and negative $\mathbf{n}$ is in the negative vertical direction (down); at resonance ($\delta = 0$) it is horizontal, and for large positive δ it points up. "Large" means $\gg |\varepsilon|$; i.e. the strength ε of a resonance is also a measure of its width.

3.2. *Passage through resonance*

If δ varies slowly (adiabatically) from negative to positive, the spin reverses — the same phenomenon that is known from NMR. The condition for adiabaticity has been quantified by Froissart and Stora; they show that if in the acceleration process δ increases at the rate of $2\pi\alpha$ per turn, the ratio of final to initial polarization (z-component of spin) is

$$\frac{P_f}{P_i} = 2\exp\left(-\frac{\pi\varepsilon^2}{2\alpha}\right) - 1 \tag{23}$$

which ranges from +1 for fast traversal of the resonance to −1 for slow (adiabatic) traversal.

$2\pi\alpha$ = rate of change of $|\delta|$ per turn.
Strong resonance or slow traversal ⇒ Spin flip
Weak resonance, fast traversal ⇒ Spin unchanged
Intermediate case ⇒ Depolarization

3.3. *Mitigation*

With many resonances, depolarization is inevitable unless something is done.

1. Make all resonances so weak that they do nothing. This may be done — by exquisite orbit correction schemes — for *imperfection* resonances. It worked at the AGS (with excruciating orbit correction work). But the beam inevitably performs betatron oscillations, and therefore *intrinsic* resonances cannot be avoided.
2. As an intrinsic resonance $kP \pm \nu_z$ is approached, rapidly jump the value of ν_z, so one jumps across the resonance. This requires rapidly pulsed quadrupoles; the scheme worked at the AGS.
3. MAKE THE RESONANCES VANISH!
   Derbenev and Kondratenko (Novosibirsk, ~1976) invented "Siberian snakes". A snake is a device that rotates the spin by 180° about a horizontal axis (longitudinal or radial or in between) without perturbing the orbit. If the snake is at an angle φ from longitudinal its spin matrix is simply

$$\exp\left(\frac{i}{2}\pi(\sigma_1\cos\varphi + \sigma_2\sin\varphi)\right) = i(\sigma_1\cos\varphi + \sigma_2\sin\varphi)\,.$$

   If we insert this in a circular ring, the one-turn matrix from a point just opposite the snake, is

$$\exp\left(\frac{i}{2}\pi\gamma G\sigma_3\right)(\sigma_1\cos\varphi + \sigma_2\sin\varphi)\exp\left(\frac{i}{2}\pi\gamma G\sigma_3\right) = i(\sigma_1\cos\varphi + \sigma_2\sin\varphi)$$

   where we have used the anticommutation property of the spin matrices. The invariant spin vector at the reference point is thus just equal to the rotation axis of the snake, and the spin tune is just 1/2, independent of the energy! (Elsewhere the spin tune is still 1/2, but the spin rotates in the horizontal plane with frequency γG).

Therefore

There is no energy at which the spin tune resonates with imperfections or oscillations.

So: Simply include a Siberian snake, and there will not be any resonances; the polarization will remain through the acceleration cycle.

It's not that simple!

When depolarizing fields are present the polarization is still affected. The snake tends to produce cancellation between errors in one turn and at the same place in the next turn, but in between the depolarizing effects can build up. So the resonances cannot be allowed to be too strong.

One can improve the situation by introducing two snakes, 180° apart, with rotation axes in the horizontal plane perpendicular to each other, say longitudinal and radial, or ±45° from longitudinal. Then the cancellation of effects takes place twice as fast, and the snakes can cope with stronger depolarizing fields. Furthermore the two-snake configuration has the advantage that the invariant spin direction in the two arcs is vertical throughout one arc (and then vertical in the opposite direction for the other), independent of energy. Stronger resonances can, in principle, be handled by more than two pairs of snakes.

How does one make these snakes? Take advantage of the fact that the spin rotation in traversing a magnetic field is γG times the bending of the orbit. At high energies, therefore, one can get a substantial spin rotation while the orbit is only distorted a little bit. A sequence of bending magnets with alternate horizontal and vertical fields (all transverse to the main orbit) will rotate the spin alternately about vertical and horizontal axes; and one may devise configurations that produce zero net orbit deflection and at the same a spin rotation through the desired 180°. What is more, the needed magnetic fields are the same at all energies, because the spin rotation in going through a distance L in a given transverse magnet is just

$$\frac{\gamma GBL}{B\rho}$$

and, since the magnetic rigidity $B\rho$ is proportional to energy (relativistically), this is independent of energy. But the orbit deflection is inversely proportional to $B\rho$, and therefore can be quite large at low energies.

Many different configurations of sequences of magnets have been devised which are "snakes" (180° spin rotators) with zero net orbit deflection. All versions require a total of the order of 20 to 30 Tesla-meters of magnets per snake.The design criteria, depending on the partiular application desired, are:

(i) Specify the rotation axis (longitudinal, transverse radial, or some angle in between).
(ii) Make the snake as compact (short) as possible so as to fit into a given insertion in a storage ring or accelerator.
(iii) Minimize the peak orbit excursion inside the snake.

3.4. *AGS resonances*

Such snakes are not feasible for the AGS because its low injection energy (1.5 GeV), where the orbit excursions inside the snake would be excessive; besides, the longest straight section in the AGS is 3 m long, so there is not enough room. However, *partial* snakes (rotating by much less than 180°) turn out to be very useful.

In AGS, insert a single "partial snake": a solenoid that just fits into one of the 3 m straight sections, rotating the spin by 3–5% of a full flip. This is equivalent to moderately strong imperfection resonance at every integer. These are now all strong enough so that the spin reverses completely as each integral γG is passed: imperfection resonances no longer depolarize.

Similarly the intrinsic resonances can be made strong enough for complete spin flip by exciting coherent oscillations of sufficient amplitude with an rf deflecting magnet; the excitation is slowly switched on just below the resonance, and then off slowly just above.

3.5. *RHIC*

For RHIC we have settled on a design that uses four helical magnets, because this design gives the smallest orbit excursion (3 cm maximum at RHIC injection energy). Each magnet has a field

$$B_x = B_0 \sin ks, B_z = B_0 \cos ks; 0 < ks < 2\pi\,;$$

i.e. a full 360° helix, 2.4 meters long. The whole snake is 10.56 m long and fits into the space between quadrupoles 7 and 8 in the RHIC lattice. Each RHIC ring has two such snakes, with rotation axes 45° and −45°.

With these snakes, tracking simulation programs show that the spin can indeed be preserved (at least to 80–90%) through the whole acceleration cycle up to 250 GeV, but at those energies where without snakes there would be strong resonances (strength ε about 0.4 to 0.6 units) one gets strong depolarizations from which the beam recovers as the dangerous energy is passed. But for best results the closed orbit errors have to be reduced — by some orbit correction schemes — to rms displacements not much more than about 0.2 mm.

In these schemes the spin is vertical except inside the snakes. For many, if not most experiments on spin physics one may want to investigate helicity states with the spins of the colliding particles parallel or antiparallel to their velocities. To make this possible we need further rotators, this time through 90°. The most advantageous rotator designs are again with four helical magnets each; one rotator (vertical to longitudinal spin) upstream of STAR, a second one restoring the spin back to vertical downstream; and a similar pair upstream and downstream from PHENIX.

So the snakes and rotators and all other hardware are now under construction. Next year, or in 2001, we expect to begin working with polarized beams in RHIC.

3.6. *HERA*

It may also be possible to work with polarized protons at the electron–proton collider HERA in Hamburg. HERA already works with polarized electrons or positrons. Studies are now underway to see whether it is feasible to do polarized protons as well.

The problem is more difficult than in RHIC, for several reasons: Resonances tend to be stronger at higher energy. Therefore two snakes may not be sufficient. HERA has essentially four-fold periodicity; therefore four snakes can be inserted.

What is worse, the HERA proton ring is not exactly planar. To provide for separation between electron and proton orbits, the ring has been designed with several sections where the beam is pushed up vertically and then down again. This greatly complicates the spin dynamics, because if we have a vertical bump, then some horizontal bending is seen, and then a bump down, the spin effects of the two bumps do not cancel. Anferov (Michigan) has devised special "flattening snakes" which can neutralize this effect, but that is more hardware.

Studies are going on at Michigan as well as at DESY, and it seems — but is not yet certain — that a combination of four "ordinary" snakes plus four "flatteners" can preserve polarization — provided the orbit errors are very well corrected (to ± 0.2 mm excursion rms).

In conclusion, polarized protons in the multi-100 GeV range are under way at RHIC and maybe at HERA, and promise new insights into the behavior of particles in different spin states.

Li Hua Yu

RESEARCH AND DEVELOPMENT TOWARDS X-RAY FREE ELECTRON LASERS

LI HUA YU

Brookhaven National Laboratory, Upton, NY 11973, USA

In this paper we describe the basic principles of the high gain free electron lasers (FEL) and the research and development towards X-ray FEL. We discuss several different schemes of seeded single pass FEL to improve the temporal coherence, and the advantages and limitations of these schemes. We discuss the High Gain Harmonic Generation (HGHG) experiment in the infrared region, and the recent Deep Ultraviolet FEL experiment using the HGHG method at Brookhaven National Laboratory.

It is a great honor for me to report my work at this symposium in honor of Professor C. N. Yang's retirement.

In the early seventies, I was teaching high school in a People's Commune in China. I dreamed that one day I would study particle physics as a student of Prof. Yang. At that time, this did not seem like a possibility. However, my dream remained alive as I learned the philosophy that everything changes, and nothing is certain except for change itself.

Things indeed began changing rapidly in the late seventies. In 1979, the dream suddenly became a reality as I found myself sitting with Prof. Yang discussing physics! Under his guidance, I began to work on the problem of monopoles. During our discussions, I learned about the many different aspects of science. I realized that different fields of science are alive; they have life cycles of their own. There are periods of rapid growth as well as slowing of development. I began to realize that the philosophy about change applies to the field of science as well. Prof. Yang taught me that research is like searching for gold; it is very difficult to find gold in a place that has been searched thoroughly. However, if you look for gold in uncharted territory, the chances of finding gold are much higher. One day, after a seminar about atomic physics, Prof. Yang told me I should open my mind like an antenna to search for and capture new information. From then on, I started to pay attention to seminars in a different way and always kept in mind to look for active new fields of interest. I found the burgeoning field of free electron lasers to be both interesting and filled with potential, and henceforth followed this direction.

Since then, there have been significant changes. The new field of high gain free electron laser was developed, and the concept of the X-ray free electron laser had evolved. Here, I would like to discuss the progress that has been made in the field as well as some of my work in this field.

1. The Basic Principles of FEL

In 1976, John Madey and associates demonstrated the first FEL with an experiment in the infrared region.[1] This FEL operated in a regime we now call the small gain regime. Its basic principle is shown in Fig. 1. In this device, an electron beam, collinear with an input laser, goes into a magnet with an alternating magnetic field called the undulator. The beam travels along the axis of the undulator. The simple following argument will explain that if the electron beam energy is appropriate, the laser light will be amplified. Let us consider a reference frame moving in the same direction as the electron beam. In this frame, due to the Doppler effect, the wavelength of the laser light will be longer. For the same reason, the wavelength of the undulator will be shorter because it is moving in the opposite direction as the reference frame. If we choose the speed of the reference frame appropriately, these two wavelengths will become equal. We then define the reference frame as a "resonance frame". In this frame, the two electro-magnetic waves traveling in opposite directions will form a standing wave, with node points where the field is minimal and peak points where the field is maximal, as shown in Fig. 1. Based on the basic electro-magnetic theory, we know that the electrons oscillating in this standing wave, experience a "pondermotive force" due to their oscillating speed

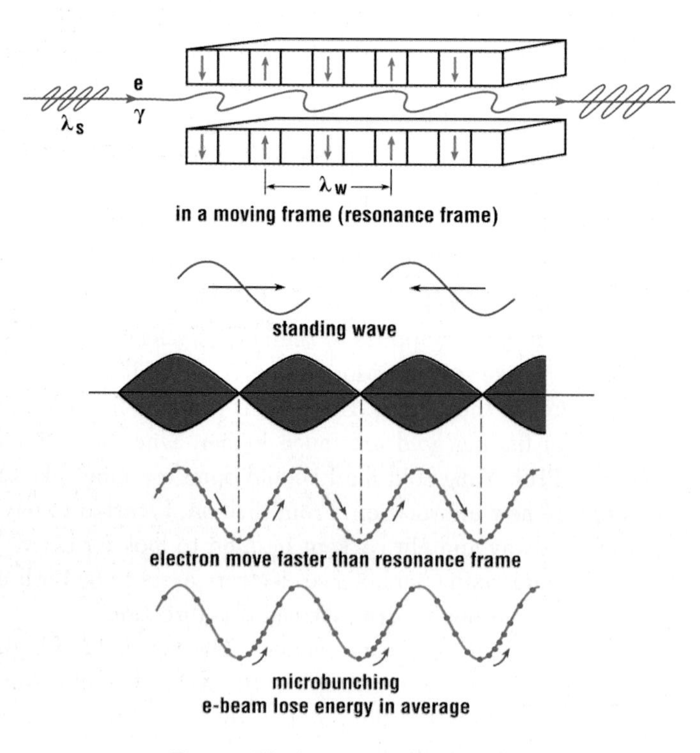

Fig. 1. The basic priniciple of FEL.

and the alternating magnetic field in the standing wave, and tend to move into the node points. This "ponderomotive force" can be described by a sinusoidal potential well, as shown in Fig. 1. Consider the case when the electrons are not standing still in the resonance frame, but are moving in the forward direction slowly. They are arranged like an initially uniform distribution of many bicycles riding in a sinusoidal mountainous area, where the ones moving down the slope speed up while those moving up the slope slow down, as indicated in the figure. After a while, the electrons will become clustered near the valleys of the mountain (the node points of the ponderomotive potential well). We call this clustering "micro bunching" because in the laboratory frame, the spacing of the clusters is equal to the optical wavelength of the input laser. However, because the electrons are moving forward in the resonance frame, the micro bunching occur not exactly in the node points of the standing wave but are slightly to the right of those node points, as shown in the figure. Thus the picture becomes asymmetrical, as more electrons are moving up the slope than down the slope. It becomes clear that more electrons are losing energy than gaining. Hence, on average, the electrons are losing energy when all the electrons are taken into account in the laboratory frame. Due to energy conservation, this lost energy must be compensated by an increase of the laser energy, thus the input laser is amplified as it exits the undulator. This is the basic principle of the free electron laser (FEL).

It is clear that the concept of "micro-bunching" plays a crucial roll in this process of amplification. The radiation from the different micro-bunches, separated by an integral numbers of optical wavelengths, are coherent with each other, as compared with the incoherent radiation from individual electrons when they are not micro-bunched. This coherence greatly enhances the output radiation, just like "super-radiation" in atomic physics.

The first free electron laser worked in the "small gain regime", where the laser is amplified by a few percent after passing through the undulator. Later on, the "high gain regime" was developed, where the gain is in the millions.

In the small gain regime, the FEL uses an "oscillator" configuration to achieve a high output power, as shown in Fig. 2. In this configuration, the electron beam consists of a train of micro-pulses that forms a "macro-pulse". As an example, these trains of pulses are provided by a linac. In Fig. 2, we show a typical set of parameters. The macro-pulses are separated by tens or hundreds of milliseconds, while the trains typically have lengths of a few microseconds. Each micro-pulse is a few pico-seconds long and the pulses are separated by a few tens of nano-seconds. In case of the oscillator configuration, as shown in Fig. 2, the first electron micro-pulse in the undulator generates a spontaneous radiation pulse, which is reflected back by a mirror that is to the right of the undulator. After traveling back through the undulator in the opposite direction, the pulse is reflected by the mirror that is to left of the undulator. When it goes through the undulator again, it is arranged that the second micro-pulse of the macro-pulse train overlaps with and amplifies the radiation. The radiation is reflected back by the right mirror again, and the process

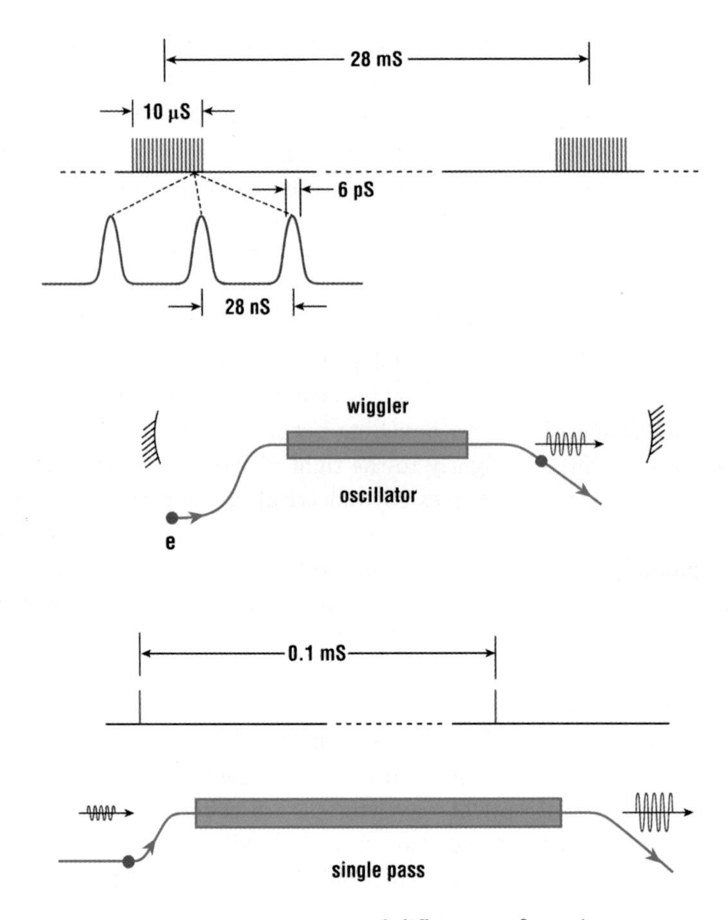

Fig. 2. Time structure and different configurations.

will repeat itself by the third micro-pulse, etc. After many repetitions of this process, when tens or hundreds of the micro-pulses forming the macro-pulse pass through the undulator, the radiation is amplified millions of times. Eventually the radiation intensity builds up so high, that the energy loss of the electrons within one pass through the undulator becomes very large. When this occurs, the electrons become out of resonance with the input radiation wavelength. Hence the gain continues to reduce until it is equal to the losses in the cavity formed by the two mirrors and the space between them. This is the basic principle of the oscillator configuration, which usually operates in the small gain regime.

2. High Gain FEL, SASE, and Gain Length

One of the disadvantages of the oscillator configuration is that there are no mirrors with good reflectivity available for wavelengths shorter than Deep UV (below one-hundred nanometers), let alone within the X-ray range. Thus the concept of a new configuration of "high gain FEL" was developed in the early eighties. As explained before, in an FEL, the radiation produces micro-bunches in the electron beam, and in turn, the micro-bunches amplify the radiation because they are coherently in phase with the input radiation. Because this is a positive feedback mechanism, if the quality of the electron beam is very good and the undulator is sufficiently long, there is an exponential growth of radiation intensity.

When there is no input in the FEL, the spontaneous radiation produced by the electrons at the beginning of the undulator is amplified exponentially. This is called the "Self Amplified Spontaneous Emission" (SASE).[2,3] If the length of the undulator is long enough and the SASE gain is high enough, the energy loss of the electrons within a single pass through the undulator becomes very large. When the loss is large enough, the electrons become out of resonance with the input radiation, hence the exponential growth stops before the beam reaches the end of the undulator. This is referred to as saturation. The main advantage of the SASE FEL is that there is no need for mirrors, making it possible to generate intense hard X-rays by FEL.

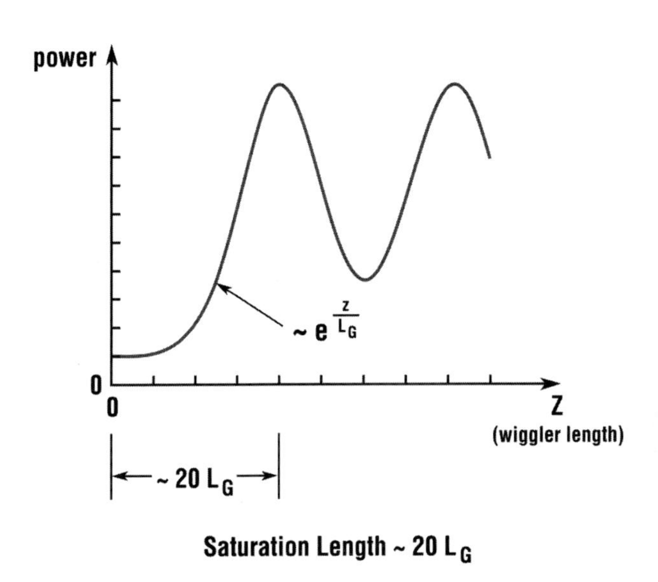

Fig. 3. Gain length and saturation.

The power gain length L_G is the distance in the undulator along which the radiation power increases by a factor of e. Theory shows it takes approximately 20 L_G of the undulator for the SASE power to grow by a factor of about 10^8 and reach saturation (see Fig. 3). Thus whether it is possible to achieve hard X-ray FEL depends on whether it is possible to make 20 L_G a realistic distance in practice. This was made possible by advances on high gain FEL theory and high brightness electron beam in the middle to later half of the 1980s.

3. Relations Between the Gain Length and the Electron Beam Quality

The earlier works on high gain FEL were for idealized cases. The theory was one-dimensional, so the diffraction loss was neglected. The electrons were all assumed to be parallel to the axis of the undulator, so the angular spread was neglected.

Because of angular spread and energy spread, different electrons have different longitudinal velocities. This causes the micro-bunches formed during the FEL process to disperse. As shown in Fig. 4, the electrons in resonance with the radiation and in phase with the wave front will remain in phase after one undulator period, but the electrons with different longitudinal velocity will be out of phase with the electrons in resonance, causing the micro-bunches to disperse. The figure illustrates that within one period, one electron is out of phase with the resonant electron by $\Delta\lambda$. If the accumulated de-bunching in one gain length is comparable to a quarter of a wavelength, the micro-bunching generated by the exponential growth cannot compete with the de-bunching process, and the gain will be significantly reduced. Hence angular spread and energy spread are very important qualities of the electron beam.

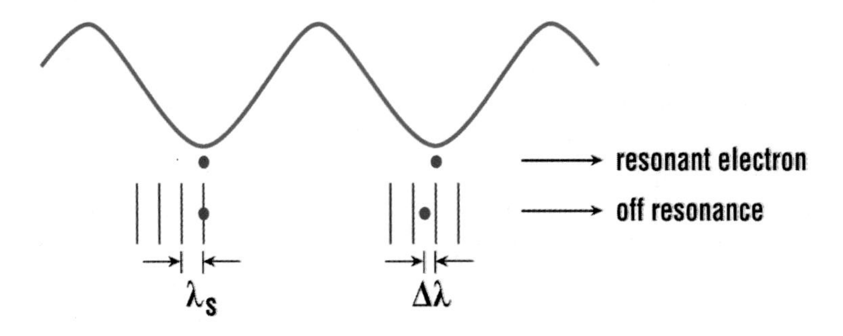

Fig. 4. Energy spread and angular spread degrade micro-bunching.

To achieve high gain, we need to focus the electron beam in the undulator to maintain small beam size so that the current density is high. Roughly speaking, the product of the beam size and the angular spread of the electron beam equals the transverse phase space area. Because this area is constant, the stronger focusing will increase the angular spread. This phase space area is an important measure of the quality of the electron beam. In accelerator terminology, it is defined as the emittance ϵ. During acceleration, the product of the emittance ϵ and energy γ remains constant, hence the quality of the electron beam is characterized by this product and defined as the normalized emittance $\epsilon_n = \epsilon\gamma$.

Theory shows that to prevent the angular spread from significantly reducing the gain given by the one-dimensional theory, emittance should be much smaller than the wavelength. Thus to achieve hard X-rays, the emittance needs to be much smaller than an Angstrom. The quality of the electron beam, as characterized by the normalized emittance, the energy spread, and the peak current, is crucial in determining how short the wavelength of a high gain FEL can be.

Much work was done in the mid eighties to calculate the effect of diffraction, focusing, and electron beam quality on FEL gain.[4] The result can be summarized as a scaling function.[5] This work showed that the gain length can be calculated in a scaled form as a function of emittance, energy spread, and focusing strength in the undulator. That is, all these parameters are expressed as their ratio to a scaling constant, which is determined by the electron beam energy and current. The scaling function shows a favorable scaling toward shorter wavelengths. The required parameters are scaled with a weak dependence on the wavelength, i.e. on its square root. For a given high gain free electron laser at a wavelength λ, let us consider the parameters necessary to reduce the wavelength and to keep the same growth rate per undulator period. We find that the electron energy must be increased by $\gamma \propto 1/\sqrt{\lambda}$, the electron beam current increased by $I_0 \propto 1/\sqrt{\lambda}$, the normalized emittance reduced by $\varepsilon_n \propto \sqrt{\lambda}$, and the relative energy spread σ_γ/γ to remain constant. This relatively weak dependence on wavelength indicated the feasibility of X-ray FEL. The scaling function provides rapid calculation of the gain length to help determine the quality of the electron beam required for hard X-ray FEL.

Another important advance in the mid-eighties is the development of a new high brightness electron source: the photo-cathode radio frequency electron gun.[6] In this device, a short laser pulse of a few pico-seconds is injected into a cathode within a radio frequency (RF) cavity, in which the photo-electrons from the cathode are accelerated by the high RF field. Without the high field in the cavity, the space charge of the electron pulse would blow up the electron beam size, hence increasing the emittance. With the high field, the electrons are accelerated to relativistic speed in a very short period of time. The space charge effect is largely reduced once the electrons are highly relativistic, thus achieving a much lower emittance.

To increase the peak current, the electron pulse length can be reduced through a process called magnetic compression. For the sake of brevity, we will not elaborate on this process here. We point out the fact that the product of electron pulse length and the energy spread remains constant during the compression because of the phase space conservation. Thus the energy spread σ_γ is increased when the electron pulse length is reduced. This appears to be harmful for FEL. However, according to the scaling gain function we mentioned before, the scaling parameter is σ_γ/γ, rather than σ_γ itself when the wavelength λ is reduced. Since the energy γ should be increased as $\gamma \propto 1/\sqrt{\lambda}$, and the electron beam current should be increased as $I_0 \propto 1/\sqrt{\lambda}$, it is easy to see that σ_γ/γ can be maintained constant. Thus, the compression happens to match the FEL gain scaling relation.

4. Proposals of X-Ray FELs, the Experimental Confirmation of the SASE Theory and the Lack of Temporal Coherence in SASE

Since the advent of the photo-cathode RF gun technology, the electron beam quality has been steadily and dramatically improving since the early 1990s. This progress, combined with those in the high gain FEL theory, makes it possible to consider the development of X-ray FEL. There was a series of workshops[7] organized to identify crucial directions of development and a possible parameter range.

Following this period, Stanford Linear Collider (SLAC) proposed the Linear Collider Light Source (LCLS) project for an FEL at 1.5 Angstrom using the existing linac.[8] According to the proposal, the linac energy is 14 GeV and the required electron beam normalized emittance is 1.5 mm-mrad, with a peak current of 3400 ampere. For this electron beam and an undulator period of 3 cm, the gain length is 5.6 m and the required total undulator length is 120 m. The output radiation peak power is 10 GW, which is eight orders of magnitude higher than what the undulator radiation can produce at the present.

In Europe, DESY also proposed a TESLA VUV FEL project aimed at a wavelength of 60 Angstroms.[9] In this proposal, the accelerator is a 1 GeV superconducting linac, where the required e-beam has a peak current of 2500 ampere with a normalized emittance of 2 mm-mrad. For an undulator period of 2.73 cm, the gain length is 1.3 m, and the total undulator length is 26 m. The output power is expected to be 2 GW.

However, the output of the SASE scheme is not temporally coherent, i.e. the coherence length of the output pulse is much shorter than the pulse length itself. As shown in Fig. 5, the output pulse consists of a train of pulses, each of which originates from the amplified spontaneous radiation of one electron. Since the phases of different electrons are random, the coherence length of the output train of pulses is determined by the length of each pulse. Since the wave front of the radiation of an electron moves ahead of the electron by one wavelength after each undulator period (see Fig. 4), it is ahead of the electron by $N_w\lambda$ at the end of the undulator, where λ is the radiation wavelength. As a result, the coherence length of the train is at most equal to $N_w\lambda$, the length called the slippage distance. In actuality, theory showed that the pulses formed at saturation have the coherence length less than the slippage distance by a factor of about 2.2. If the electron pulse length is $n\lambda$, which is much longer than the coherence length as shown in Fig. 5, i.e. if $n \gg N_w$, the output spectrum will consist of many spikes. Theory predicts that the number of spikes is equal to the ratio of the electron pulse length over the coherence length, i.e. the number of coherence lengths within the electron pulse. As shown in Fig. 5, the width of the spikes is $1/n$, and the width of the spectrum is $2.2/N_w$ at saturation. The bandwidth of the spontaneous radiation of an undulator is $1/N_w$, so the SASE bandwidth is even larger than the spontaneous radiation.

This is related to the large shot to shot fluctuation. Since the phases of different electrons are random, the distribution of the spikes in the spectrum is also random.

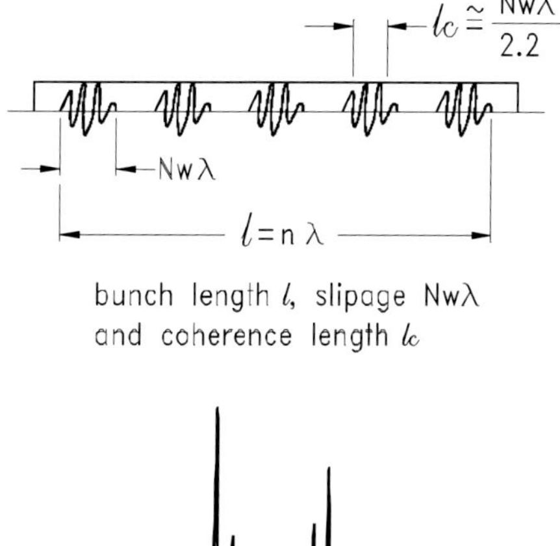

Fig. 5. SASE pulse and spectrum showing lack of temporal coherence.

Therefore, if one uses a monochromator with bandwidth of $1/n$ to achieve full coherence, its spectral window may fall on or off one of the spikes, resulting in 100% intensity fluctuation.

Since 1997, a series of SASE experiments done by different laboratories around the world confirmed the SASE theory, including the gain length, the saturation power, the bandwidth, the spikes in the spectrum, the fluctuation, etc. Among them are,

- LANL, UCLA. Collaboration, SASE at 12 μm (1998),[10]
- BNL, SASE at 0.8 μm (1998),[11]
- APS (project LUTLE), SASE saturation at 0.4 μm (2000),[12]
- BNL, SLAC, UCLA collaboration (project VISA), SASE saturation at 0.8 μm (2001),[13]
- DESY: SASE saturation at 0.1 μm (2002),[14]
- BNL (project UVFEL), SASE at 0.266 μm (2002).[15]

The success of these experiments clearly showed the rapid progress towards shorter and shorter wavelengths. The electron beam parameters from the photo-cathode RF electron gun are approaching the required value for the LCLS proposal.[8] For example, the beam parameters from the RF-gun at the Accelerator Test Facility

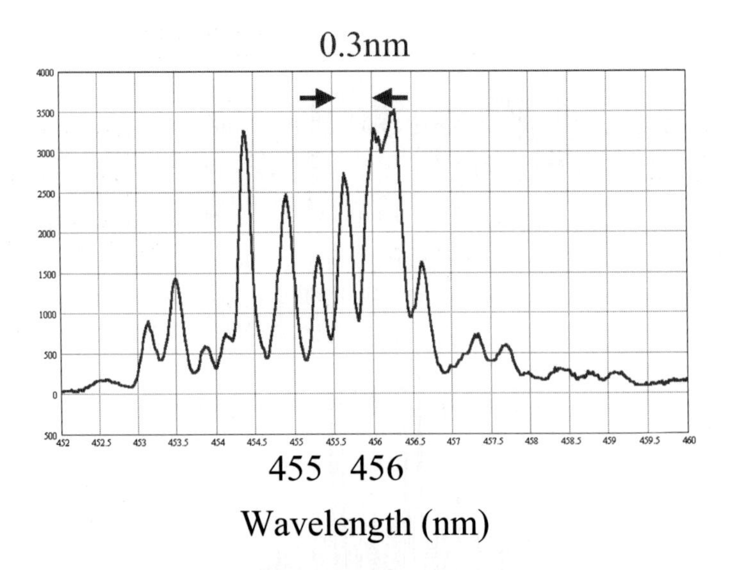

Fig. 6. SASE spectrum at 400 nm for a 4 ps electron pulse.

(ATF) of the BNL during the VISA experiment were already close to the design parameters in LCLS, the charge per pulse was 0.5 nC, with normalized emittance of 0.8 mm-mrad.[13] In the mean time, much research is being conducted to study how to preserve the quality of the electron beam during the beam transport and the previously mentioned magnetic compression process to increase peak current.

However, these experiments also confirmed the statistical properties of the SASE process: the spectrum with a bandwidth larger than the spontaneous radiation, the spikes in the spectrum, and the large intrinsic fluctuation. Figure 6 shows an example from an experiment done at BNL in April 2002, where a single shot spectrum measured during the 400 nm SASE of the UVFEL shows the random spikes. Clearly, it is desirable to develop a method to generate temporally coherent FEL output.

5. Approaches to Improve the Temporal Coherence and the Basic Principle of High Gain Harmonic Generation (HGHG)

One example of the different approaches to improve the temporal coherence is the "Two-Stage FEL" scheme proposed by DESY group.[16] As shown in Fig. 7, the SASE output from the first undulator is sent through a monochromator to reduce the bandwidth. Then it is sent into the second undulator as a seed while the electron beam is by-passed through the monochromator after it exits through the first undulator. After this, it is sent into the second undulator to amplify the narrow band seed. As was explained before, the output from the monochromator will have 100% fluctuation if the bandwidth is chosen to achieve full coherence. When the second undulator is sufficiently long, the fluctuation can be reduced. However, to

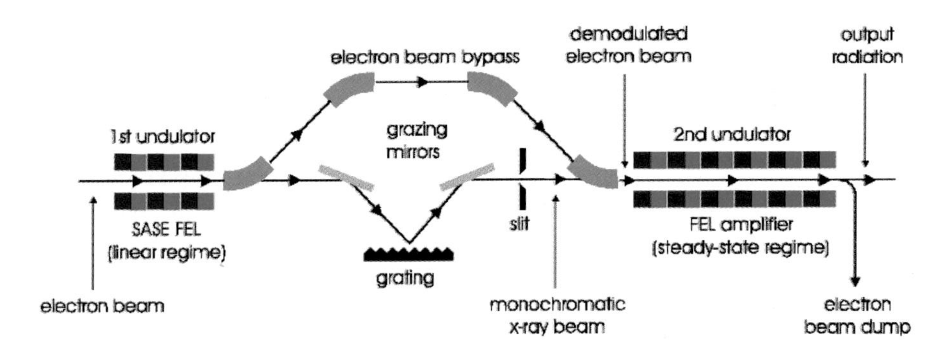

Fig. 7. Two staged FEL scheme (DESY).

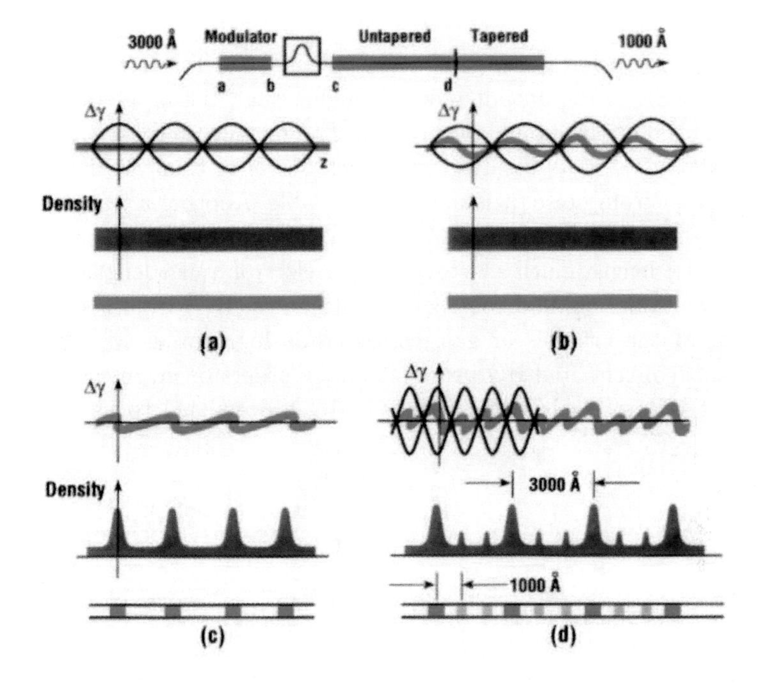

Fig. 8. The basic principle of HGHG.

prevent the SASE of the second undulator from competing with the amplified seed, its length should not exceed the saturation length. Hence, there is always a residual intrinsic fluctuation, even for a perfectly stable electron beam.

Since the late 1980s at BNL, we recognized the lack of temporal coherence in the SASE output during our theoretical study, and have begun to develop a scheme called High Gain Harmonic Generation (HGHG) to achieve short wavelength and full coherence.[17,18]

In Fig. 8, we describe the basic principle of the HGHG FEL. At point "a" on the top of the figure, the initial time and energy phase space distribution, the electron density profile and its density distribution are all uniform, as given in Fig. 8(a). A small energy modulation is then imposed on the electron beam (see Fig. 8(b)) by its interaction with a seed laser in a short undulator (the modulator). The energy modulation is converted to a coherent spatial density modulation (see Fig. 8(c)), as the electron beam traverses a dispersion magnet (a three-dipole chicane, shown as the section between (b) and (c) on the top part of Fig. 8). A second undulator (the radiator), tuned to a higher harmonic of the seed frequency (ω), causes the micro-bunched electron beam to emit coherent radiation at the harmonic frequency ($n\omega$), followed by exponential amplification (see Fig. 8(d)) until saturation is achieved. The HGHG output radiation has a single phase determined by the seed laser and its spectral bandwidth is Fourier transform limited. In Fig. 9, we schematically show the initial coherent radiation and then the exponential growth and saturation in the radiator.

This process is a harmonic generation process, and the harmonic is amplified by high gain FEL so the output power is higher than the input power, hence the name of High Gain Harmonic Generation. The output of HGHG has nearly full temporal coherence, so its bandwidth is much narrower than SASE. Also, because it does not start from noise, it is much more stable. Another advantage is that the output pulse length is determined by the seed laser, thus it is possible to generate an output pulse length much shorter than the electron pulse length.

The conventional method to generate harmonics (HG) from laser is to use the nonlinearity of the crystals or gas, which needs high intensity input. Since the output of this conventional HG process is many orders of magnitude smaller than the input, it is impossible to cascade the HG process, i.e. to use the output of

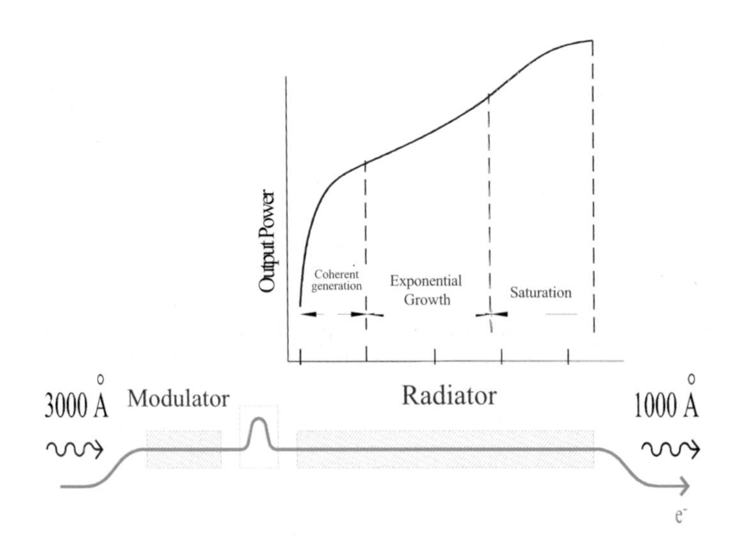

Fig. 9. Power (logarithmic scale) as function of distance in the radiator.

one HG stage as the input of next HG stage to reach much shorter wavelength. However, because the output power of the HGHG is on the same order as or much larger than the input power, it becomes possible to cascade several HGHG stages to reach X-ray region.[19]

These advantages make the HGHG an attractive direction of development towards X-ray FEL. In the late 1980s, this concept was at its beginning, the e-beam quality from the photo-cathode RF electron gun was still not adequate for X-ray FEL. It was our opinion that the development towards X-ray FEL should be step-wised, i.e. we should start from longer wavelengths and gradually develop towards shorter wavelengths. In the early 1990s at BNL, we proposed to develop a Deep UVFEL program, using the HGHG process to generate intense and coherent radiation at or below 100 nm, starting from a seed at 300 nm or shorter.[17,18] Before starting this DUVFEL program, we first planned to carry out a proof-of-principle experiment, converting 10 μm CO_2 laser into high power 5 μm output by HGHG.

6. The HGHG Experiments at the BNL

The first HGHG experiment was successfully accomplished in 1999 at the Accelerator Test Facility (ATF) of the BNL. The results have been published in Science,[20] and the detailed characterization of the output coherence properties is published in PRL.[21] A schematic diagram is shown in Fig. 10, together with the photos of the main components. The radiator is 2 m long, sufficiently long for the HGHG to reach saturation. In Fig. 11, we show the measured multi-shot spectrum of the HGHG as compared with the SASE output from the same undulator under the same electron beam condition. We need to multiply the SASE power spectrum by 10^6 in order to show it on the same scale as the HGHG spectrum. To achieve the

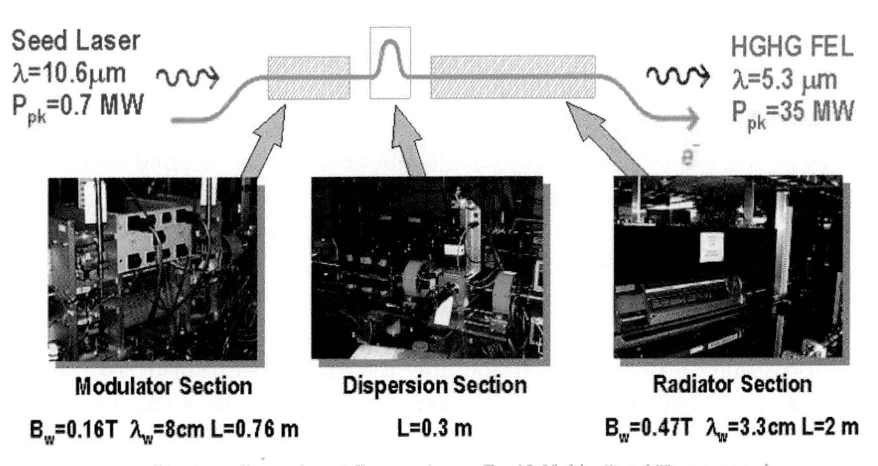

Fig. 10. The setup for the proof-of-principle experiment for HGHG.

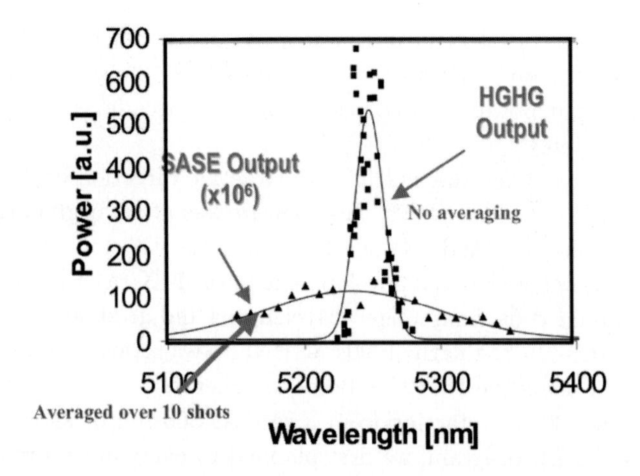

Fig. 11. Multi-shot spectrum of HGHG and SASE for the same electro beam conditions.

same output power for the SASE, the undulator should be 3 times as long, i.e. 6 m long. Because of the large fluctuation, each point for the SASE spectrum is the average of ten shots while each point for HGHG reflects a single shot. This shows that the HGHG output is much more stable than SASE. The fluctuation in HGHG output is due to the fluctuation in the electron beam parameters and the CO_2 laser power. The coherence length of the HGHG output was also found to be approximately equal to the pulse length, i.e. it is nearly fully coherent. In short, this experiment confirmed the theory and the simulation. In 2000 we started the work on the DUVFEL experiment.

The DUVFEL experiment is carried out in the Source Development Lab (SDL) of the National Synchrotron Light Source (NSLS). In February of 2002, we achieved SASE at 400 nm in the radiator undulator named NISUS.[15] In Fig. 12, the radiation power measured along the NISUS undulator is plotted versus the distance in the NISUS, showing the exponential growth of radiation. This plot gives the gain length of 0.9 m. When compared with the gain length of 1.1 m based on the design parameters, this shows that the system already satisfied the required conditions for the HGHG experiment. In May we began to install the seed laser injection line, and in September we installed the 0.8 m modulator named the MINI undulator. In late October, we achieved successful output of HGHG, generating 266 nm from an 800 nm titanium sapphire laser seed. In Fig. 13, we show the HGHG and the SASE single shot spectrum for the same electron beam condition. In order to show them on the same scale we need to multiply the SASE spectrum by 10^5. The fluctuation of the HGHG is 7%, as compared with the 41% of SASE for the same electron beam condition (Fig. 14). The fluctuation of HGHG is due to the large electron beam fluctuation on the day of the measurement, while that of the SASE came from both the intrinsic fluctuation and the beam fluctuation. This shows the significant improvement in the stability of the HGHG process.

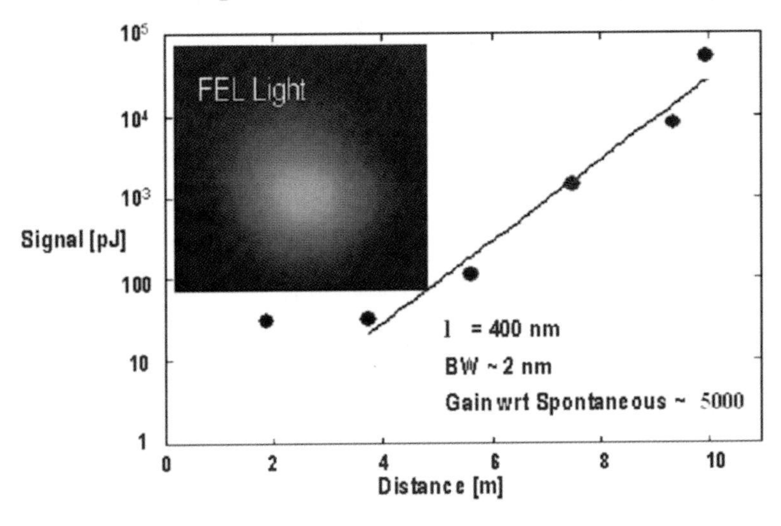

Fig. 12. The output radiation profile and the measured power as function of distance in NSUS in the 400 nm SASE experiment in February 2002.

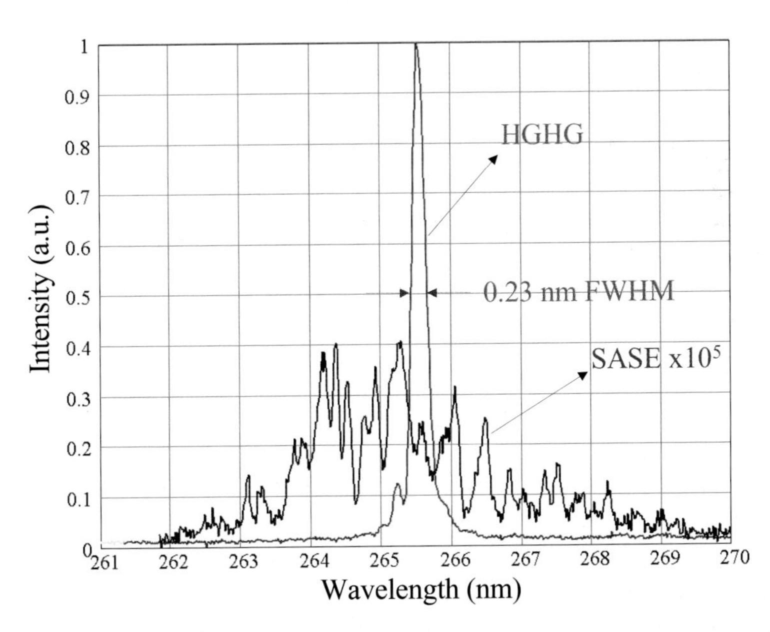

Fig. 13. The single shot spectrum of HGHG and SASE under the same beam condition on March 19, 2003.

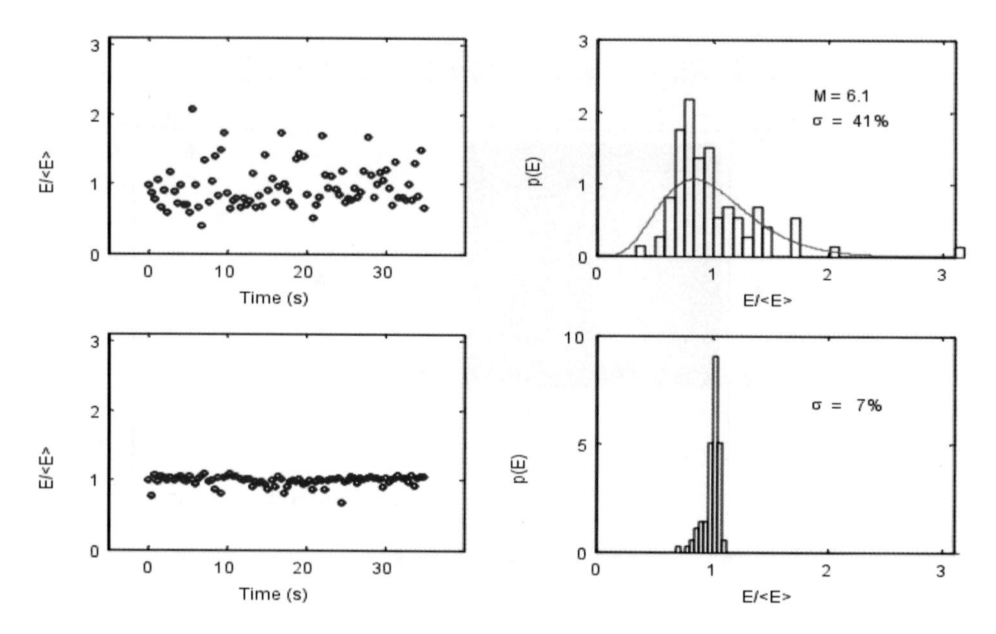

Fig. 14. Output intensity versus time and its distribution histogram for SASE (top row) and HGHG (bottom row) under the same electron beam condition.

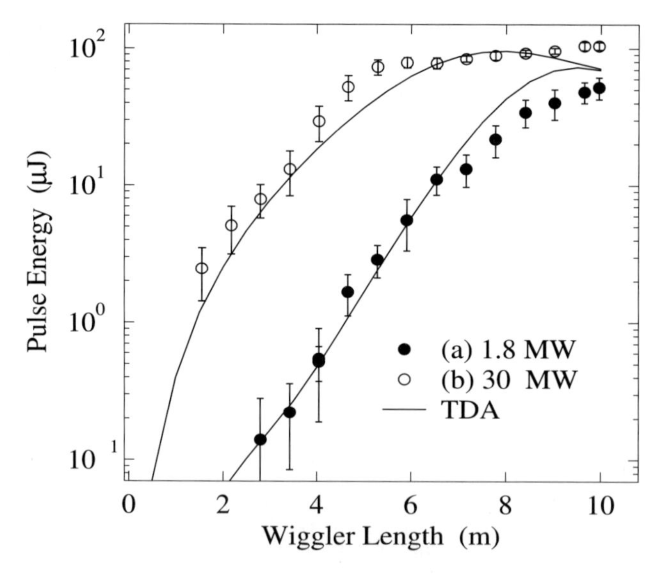

Fig. 15. Output intensity versus the undulator distance in logarithmic scale showing exponential growth, and saturation at about 5 mwhen seed power was 30 MW in case (b). TDA is a simulation.

The output reached 130 μJ with a sub-pico-second pulse length. The measure power versus undulator distance shows clearly the saturation (Fig. 15). The theory predicts a third harmonic output at 88 nm that would be 1% of the fundamental at

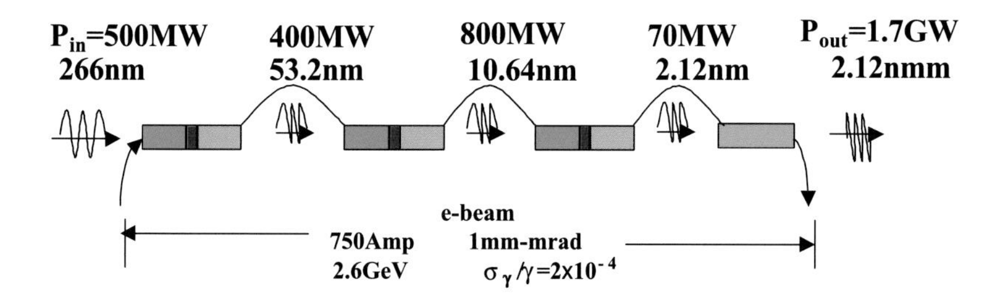

	1st Stage		2nd Stage		3rd Stage		Amplifier
λ (nm)	266	53.2	53.2	10.64	10.64	2.128	2.128
λ_w (cm)	11	6.4	6.4	4.16	4.16	2.7	2.7
L_w (m)	2	6	2	8	2	4	12
L_G (m)	1.6	1.3	1.3	1.4	1.4	1.75	1.75

$L_{\text{total}} = 36$ m to reach 1.7 GW

Fig. 16. The schematic diagram for a soft X-ray FEL based on cascading HGHG concept.

266 nm when saturation is achieved. This has been confirmed recently. The 88 nm radiation will become useful for many new types of experiments in chemistry. In January, the first user application experiment of the DUVFEL has been started. In the experiment, ethane molecules were ionized by the intense 88 nm radiation, the velocity distribution of the fragments were measured, providing information about the molecular structure and dynamics. If we upgrade our electron beam energy from the present maximum of 200 MeV to 300 MeV, we can generate coherent radiation below 100 nm with more than 100 μJ per pulse. This will produce an unprecedented high brightness radiation in this wavelength region, which will allow for many new applications in different branches of chemistry.

In the mean time, we are considering a cascading scheme of a soft X-ray FEL program[19] as a part of the NSLS upgrade proposal, as shown schematically in Fig. 16. In this scheme, we start from a 266 nm seed, then generate radiation at 53.2, 10.6 and 2.1 nm, respectively, through three HGHG stages, and finally amplify the 2.1 nm radiation to saturation at an output power of 1.7 GW. One feature of this scheme is that it can produce output pulses as short as 20 fs, which will allow for many important applications in a wide range of fields. We will not elaborate on the scheme here. Of note in the table are the input and output wavelengths of the three HGHG stages, their undulator periods, the gain lengths, and lengths of different sections, which add up to a total undulator length of 36 m. One of the technical challenges for this soft X-ray FEL is the time jitter between the laser pulses and the electron pulses. Since the electron pulse is about 500 fs and we need to use the different parts of this pulse several times during the cascade of several HGHG stages, we should reduce the time jitter to below 100 fs. At present, the time jitter without feedback system is 500 fs. With a feedback system, the desired goal is possible, even though there is still a significant challenge. To overcome this challenge will be one of the important steps towards our final goal: the temporally coherent X-ray FEL.

Acknowledgments

This work was carried out with the support of DOE Contract DEAC No. DE-AC02-98CH10886 and AFOSR/ONR MFEL Program #NMIPR01520375.

References

1. L. R. Elias, W. M. Fairbank, J. M. J. Madey, H. A. Schwettman and T. Smith, *PRL* **36**, 717 (1976).
2. Y. S. Debenev, A. M. Kondratenko and E. L. Saldin, *Nucl. Instrum. Meth. Phys. Res.* **A193**, 415 (1982).
3. R. Bonifacio, C. Pellegrini and L. Narducci, *Opt. Commun.* **50**, 373 (1984).
4. The references on this subject can be found in, for example, G. T. Moore, *Nucl. Instrum. Meth.* **A239** (1985) 19; E. T. Scharlemann, A. M. Sessler and J. S. Wutele, *PRL* **54**, 1925 (1985); J. M. Wang and L. H. Yu, *Nucl. Instrum. Meth. Phys. Res.* **A250**, 484 (1986); K. J. Kim, *Nucl. Instrum. Meth. Phys. Res.* **A250**, 396 (1986); K. J. Kim, *Phys. Rev. Lett.* **57**, 1871 (1986); S. Krinsky and L. H. Yu, *Phys. Rev.* **A35**, 3406 (1987).
5. L. H. Yu, S. Krinsky and R.L. Gluckstern, *Phys. Rev. Lett.* **64**, 3011 (1990); An earlier qualitative discussion on the scaling can be found in: C. Pellegrini, *Nucl. Instrum. Meth.* **A272**, 364 (1988).
6. J. S. Fraser *et al.*, *Proc. 1987 IEEE Particle Accerator Conf.*, Cat. No. 87CH.2387-9 (March 1987) p. 1705.
7. J. Gallardo (ed.), *Proc. Workshop Prospect for a 1 Å Free-Electron-Laser*, Sag Harbor, NY, April 22–27, 1990, BNL 52273; M. Cornacchia and H. Winick (eds.), *Workshop on Fourth Generation Light Sources*, SRRL report 92/02, Feb 24–27, 1992; I. Ben-Zvi and H. Winick (eds.), *Report on the workshop Towards Short Wavelength Free Electron Lasers*, Brookhaven National Laboratory, May 21–22, 1993, BNL Report 49651; W. Spicer, J. Arthur and H. Winick (eds.), *Workshop on Scientific Applications of Short Wavelength Coherent Light Sources*, SLAC Report 414, 1992; J. Arthur, G. Materlik and H. Winick (eds.), *Workshop on Scientific Applications of Coherent X-rays*, SLAC Report 437, 1994.
8. *Linac Coherent Light Source (LCLS) Design Study Report*, The LCLS Design Study Group, Stanford Linear Accelerator Center (SLAC) Report No. SLAC-R-521, 1998.
9. *A VUV Free Electron Laser at the TESLA Test Facility at DESY: Conceptual Design Report*, DESY Print, June 1995, TESLA-FEL 95-03.
10. M. J. Hogan *et al.*, *Phys. Rev. Lett.* **22**, 4867 (1998).
11. M. Babzien *et al.*, *Phys. Rev.* **E57**, 6093 (1998).
12. S. V. Milton *et al.*, *Phys. Rev. Lett.* **85**, 988 (2000).
13. A. Murok *et al.*, *Proc. 2001 Particle Accelerator Conf.*, Chicago, p. 2748; The ATF beam parameters can be found in: V. Yakimenko *et al.*, *Nucl. Instrum. Meth. Phys. Res.* **A483**, 277 (2002).
14. V. Ayvazyan *et al.*, *PRL* **88**, 10, 104802 (2002).
15. A. Doyuran *et al.*, to be published in *Nucl. Instrum. Meth.* (2002).
16. J. Felhaus *et al.*, *Opt. Commun.* **140** (1997) 341.

17. I. Ben-Zvi, L. F. Di Mauro, S. Krinsky, M. White and L. H. Yu, *Nucl. Instrum. Meth. Phys. Res.* **A304**, 151 (1991).
18. L. H. Yu, *Phys. Rev.* **A44**, 5178 (1991).
19. J. Wu, and H. Y. Li, *High Gain Harmonic Generation X-ray FEL*, 2716, *Proc. PAC2001*, Chicago (2001).
20. L. H. Yu *et al.*, *Science* **289** (2000) 932.
21. A. Doyuran *et al.*, *Phys. Rev. Lett.* **86**, 5902 (2001).

Tai Tsun Wu

REMARKS ON YANG–MILLS THEORY

TAI TSUN WU*

Gordon McKay Laboratory, Harvard University, Cambridge, MA 02138, USA
and
Theory Division, CERN, CH-1211 Geneva 23, Switzerland

The Yang–Mills theory is the most important development in physics in the second half of this century. Some of the highlights are briefly summarized.

During my first year as a graduate student at Harvard, Professor Yang came to give a Monday Physics Colloquium. It was on the newly conceived Yang–Mills theory. Since I had not yet learned quantum field theory, I was not able to follow his discussions. It is my understanding that in 1954 Professor Yang gave just two seminars on this new theory, the first one at the Institute for Advanced Study, and this was the second one.

If I remember correctly, Professor Yang gave this Colloquium in May of 1954. If this is correct, then it was almost exactly forty-five years ago. It was the first time I met Professor Yang.

Shortly after carrying out this work, Yang and Mills gave a summary at the Washington meeting of the American Physical Society[1] and also submitted their famous paper for publication in the *Physical Review*.[2] The abstract of their talk is reproduced as Fig. 1; it took place on April 30, 1954, with Professor Freeman Dyson presiding.

The title and abstract of Ref. 2 is shown in Fig. 2, and some of the basic equations from this paper are shown in Fig. 3. There are a number of remarkable features in this paper besides the fundamental concept of non-Abelian gauge fields; let me mention just the following two.

First, after the Introduction section, Gauge Transformation is discussed before the Field Equations, as seen from Fig. 3. This is the authors' way of emphasizing the importance of gauge transformation.

Secondly, near the end of this paper, just after their Fig. 1 which shows the self-interaction of the Yang–Mills b field, there is a paragraph discussing its mass. This paragraph is reproduced as Fig. 4. We shall return to this prophetic paragraph later.

*Presented at *Symmetries and Reflections*, a Symposium in honor of Professor C. N. Yang, May 22, 1999. Work supported in part by the United States Department of Energy under Grant No. DE-FG02-84ER40158.

M7. Isotopic Spin Conservation and a Generalized Gauge Invariance.* C. N. YANG† AND R. MILLS, *Brookhaven National Laboratory*—**The conservation of isotopic spin points to the existence of a fundamental invariance law similar to the conservation of electric charge. In the latter case, the electric charge serves as a source of electromagnetic field; an important concept in this case is that a gauge invariance which is closely connected with (1) the equation of motion of the electromagnetic field, (2) the existence of a current density, and (3) the possible interactions between a charged field and the electromagnetic field. We have tried to generalize this concept of gauge invariance to apply to isotopic spin conservation. It turns out that a very natural generalization is possible. The field that plays the role of the electromagnetic field is here a vector field that satisfies a nonlinear equation even in the absence of other fields. (This is because unlike the elecromagnetic field this field has an isotopic spin and consequently acts as a source of itself.) The existence of a current density is automatic, and the interaction of this field with any fields of arbitrary isotopic spin is of a definite form (except for possible terms similar to the anomalous magnetic moment interaction terms in electrodynamics).**

*Work performed under the auspices of the U. S. Atomic Energy Commission.
†On leave of absence from the Institute for Advanced Study, Princeton, New Jersey.

F. J. Dyson Presiding
April 30, 1954

Fig. 1. Reference 1 in its entirety.

Generalization to other Lie groups is straightforward. Therefore, the Yang–Mills theory gives a principle for writing down the interactions of elementary particles.

It is natural to ask the question: Which Lie groups are appropriate for physics? When Yang and Mills wrote their papers, there was no way for them to prefer one Lie group over another. This important question was answered in the 1960's by Glashow, Weinberg and Salam[3] for electroweak interactions and in 1973 by Fritzsch, Gell-Mann and Leutwyler[4] from strong interactions. The answers are $SU(2) \times U(1)$ for electroweak and $SU(3)$ for strong interactions.

Aside from this important answer about the choice of Lie groups, some of the developments, both theoretical and experimental, of the Yang–Mills theory are shown in Fig. 5. For example, twenty years after 1954, the global formulation of Yang–Mills theory was accomplished.[5] This makes conection not only with non-integrable phase factors but also with fiber bundles. The resulting translation between the gauge-field and the fiber-bundle terminologies is reproduced in Table 1. The group $SU(2)$ is readily replaced by other non-Abelian Lie groups.

Conservation of Isotopic Spin and Isotopic Gauge Invariance*

C. N. YANG† AND R. L. MILLS

Brookhaven National Laboratory, Upton, New York

(Received June 28, 1954)

It is pointed out that the usual principle of invariance under isotopic spin rotation is not consistent with the concept of localized fields. The possibility is explored of having invariance under local isotopic spin rotations. This leads to formulating a principle of isotopic gauge invariance and the existence of a b field which has the same relation to the isotopic spin that the electromagnetic field has to the electric charge. The b field satisfies nonlinear differential equations. The quanta of the b field are particles with spin unity, isotopic spin unity, and electric charge $\pm e$ or zero.

Fig. 2. The title and abstract of Ref. 2.

Table 1. Translation of terminology.

Gauge Field Terminology	Bundle Terminology
gauge (or global gauge)	principal coordinate bundle
gauge type	principal fiber bundle
gauge potential b_μ^k	connection on a principal fiber bundle
S_{ba}	transition function
phase factor Φ_{QP}	parallel displacement
field strength $F_{\mu\nu}^k$	curvature
source J_μ^k	?
electromagnetism	connection on a U(1) bundle
isotopic spin gauge field	connection on a SU(2) bundle
Dirac's monopole quantization	classification of U(1) bundle according to first Chern class
electromagnetism without monopole	connection on a trivial U(1) bundle
electromagnetism with monopole	connection on a nontrivial U(1) bundle

On the experimental side, twenty-five years after 1954, the Yang–Mills particle was observed for the first time. It was the one for strong interactions, and this feat was accomplished by the TASSO Collaboration at DESY.[6] Nearly thirty years after 1954, the Yang–Mills particle for electroweak interactions was observed for the first time by the UA(1) Collaboration at CERN.[7] These pioneering experimental observations completed the list of all the vector bosons of the standard model.[3,4]

ISOTOPIC GAUGE TRANSFORMATION

$$\psi = S\psi', \tag{1}$$

$$S(\partial_\mu - i\epsilon B'_\mu)\psi' = (\partial_\mu - i\epsilon B_\mu)\psi. \tag{2}$$

Combining (1) and (2), we obtain the isotopic gauge transformation on B_μ:

$$B'_\mu = S^{-1}B_\mu S + \frac{i}{\epsilon} S^{-1}\frac{\partial S}{\partial x_\mu}. \tag{3}$$

Define now

$$F_{\mu\nu} = \frac{\partial B_\mu}{\partial x_\nu} - \frac{\partial B_\nu}{\partial x_\mu} + i\epsilon(B_\mu B_\nu - B_\nu B_\mu). \tag{4}$$

One easily shows from (3) that

$$F'_{\mu\nu} = S^{-1}F_{\mu\nu}S. \tag{5}$$

FIELD EQUATIONS

$$\mathcal{L} = -\frac{1}{4}\mathbf{f}_{\mu\nu}\cdot\mathbf{f}_{\mu\nu} - \bar{\psi}\gamma_\mu(\partial_\mu - i\epsilon\tau\cdot\mathbf{b}_\mu)\psi - m\bar{\psi}\psi. \tag{11}$$

One obtains from this the following equations of motion:

$$\begin{aligned} \partial\mathbf{f}_{\mu\nu}/\partial x_\nu + 2\epsilon(\mathbf{b}_\nu \times \mathbf{f}_{\mu\nu}) + \mathbf{J}_\mu &= 0, \\ \gamma_\mu(\partial_\mu - i\epsilon\tau\cdot\mathbf{b}_\mu)\psi + m\psi &= 0, \end{aligned} \tag{12}$$

where

$$\mathbf{J}_\mu = i\epsilon\bar{\psi}\gamma_\mu\tau\psi. \tag{13}$$

Fig. 3. A few equations taken from Ref. 2.

We next come to the question of the mass of the b quantum, to which we do not have a satisfactory answer. One may argue that without a nucleon field the Lagrangian would contain no quantity of the dimension of a mass, and that therefore the mass of the b quantum in such a case is zero. This argument is however subject to the criticism that, like all field theories, the b field is beset with divergences, and dimensional arguments are not satisfactory.

Fig. 4. The discussion on the mass of the Yang–Mills b particle in Ref. 2.

1954 YANG-MILLS THEORY

Twenty years later

Integral formalism

non-integrable phase factor

fiber bundles

EXPERIMENT

Twenty-five years later

First experimental observation of a

Yang-Mills particle

Yang-Mills particles for strong inter.

– TASSO Collaboration at DESY

Thirty years later

Experimental observation of

Yang-Mills particles for electroweak

– UA(1) Collaboration at CERN

Fig. 5. Some of the developments of the Yang–Mills theory since 1954.

Let me return to the interesting paragraph of the Yang–Mills paper shown in Fig. 4. When Professor Yang gave the seminar on this new theory at the Institute for Advanced Study in February of 1954, Pauli repeatedly asked the question: What is the mass of this field B_μ? No definitive answer was possible at that time, and by now it is clear why. The Yang–Mills particle for strong interactions, now often called the gluon, is massless, while the Yang–Mills particles Z and W for electroweak interactions are massive, about 80–91 GeV/c^2. Therefore, both possibilities, massless and massive, can happen for Yang–Mills particles and actually do. This is the reason why the question raised by Pauli could not have been answered at that time, and also why the paragraph of Fig. 4 can only be called prophetic.

In summary, since 1954 particle physicists have been learning about and developing the Yang–Mills theory, which permeates throughout particle physics. The Yang–Mills theory is the most important and deepest development in physics for the second half of the twentieth century.

Acknowledgment

I would like to thank the Theory Division of CERN for its kind hospitality.

References

1. C. N. Yang and R. L. Mills, *Phys. Rev.* **95**, 631 (1954).
2. C. N. Yang and R. L. Mills, *Phys. Rev.* **96**, 191 (1954).
3. S. L. Glashow, *Nucl. Phys.* **22**, 579 (1961); S. Weinberg, *Phys. Rev. Lett.* **19**, 1264 (1967); A. Salam, *Proc. Eighth Nobel Symp.*, May 1968, ed. N. Svartholm (Wiley, 1968) p. 367.
4. H. Fritzsch, M. Gell-Mann and H. Leutwyler, *Phys. Lett.* **B47**, 365 (1973).
5. T. T. Wu and C. N. Yang, *Phys. Rev.* **D12**, 3845 (1975).
6. B. H. Wiik, *Proc. Int. Conf. Neutrinos, Weak Interactions and Cosmology*, 18–22 June 1979, p. 113; P. Söding, *Proc. European Society Int. Conf. High Energy Physics*, Geneva, Switzerland, 27 June–4 July, 1979, p. 271; TASSO Collaboration, R. Brandelik *et al.*, *Phys. Lett.* **B86**, 243 (1979).
7. UA(1) Collaboration, G. Arnison *et al.*, *Phys. Lett.* **B122**, 103, and **B126**, 398 (1983).

Gerard 't Hooft

THE HIDDEN INFORMATION IN THE STANDARD MODEL

GERARD 'T HOOFT
Institute for Theoretical Physics
Utrecht University, Leuvenlaan 4
3584 CC Utrecht, The Netherlands
and
Spinoza Institute, Postbox 80.195
3508 TD Utrecht, The Netherlands
E-mail: g.thooft@phys.uu.nl
Internet: http://www.phys.uu.nl/~thooft/

After noting that the Standard Model could never have been discovered without the crucial insights that lead Frank Yang to writing his seminal 1954 paper with R. Mills, we continue to marvel at the rich amount of information that appears to be hidden in this model. The specific group structure of the local gauge transformations, the representations of these groups found in the fermionic and the Higgs sector, all seem to be telling us something about Nature at ultrashort distance scales. Many researchers are trying to embed the observed system in a supersymmetric scheme. Here, we suggest a more ambitious attempt: to see if one can find some deterministic underlying theory. An attempt to pioneer such an approach is explained.

1. Introduction

Many of the advances in twentieth century theoretical physics can be attributed to the discovery and subsequent exploitation of symmetries. Not only Special and General Relativity embody the importance of realizing how a precise analysis of Nature's symmetries can enhance our understanding, but also Quantum Mechanics itself owes much of its success to the exploitation, to the extreme, of symmetries such as rotation and translation symmetries, but also conformal symmetries and many kinds of internal and accidental symmetries. It was Frank Yang's deep insights that allowed him to observe how to generalize the few local symmetry concepts that existed at his time: local coordinate transformations in general relativity and the Weyl gauge invariance of electromagnetism.

Together with Robert Mills, he showed that one can write down completely self-consistent field equations for a set of fields which are very similar to electric and magnetic fields, but which induce *transitions* of particles that pass through such fields into other, related species of particles.[1] The group of local gauge transformations, which was an *Abelian* group in the electromagnetic case, is now replaced by a non-Abelian group, which may directly be linked to groups such as that of the isospin transformations, which up to that time had been treated only as *global* symmetry groups.

The theory naturally requires many particle species to form representations of such global groups, and the fact that global, non-Abelian symmetry groups already seemed to be applicable to describe the known particle species — first $SU(2)$ and then, later, $SU(3)$, could be seen as a strong indication that these ideas could become quite meaningful and important. In spite of this, however, the Yang-Mills scheme received an amount of skepticism as if it were some fringe science, during the first 15 years.[2]

Even the work of R. Brout and F. Englert,[3] and independently of P. Higgs,[4] did little to alter that status. I think that one can present several explanations for this disinterest. One is, that investigators were still searching for Nature's fundamental principles. Could gauge invariance be a fundamental principle? Then why introduce scalar fields that turn gauge symmetry into something that seems to be much uglier? Why should one rely on Quantum Field Theory, if *non-perturbatively* this theory appeared to exhibit unacceptable small-distance behaviour, whereas the perturbative formulation appears to suffer from unacceptable infinite renormalization parameters? Presently, we do know what the solid and reliable principles are on which to construct theories: Quantum Field Theories with Yang-Mills fields in them, do possess acceptable small distance features, and gauge-invariance is an acceptable but not exhaustive starting point for a theory; by adding these gauge fields, theories can be obtained with approximate or even exact asymptotic freedom, but this does not exclude the inclusion of elementary non-gauge fields, which may be scalar or fermionic, or, in the most satisfactory schemes, we may add both.

2. The Standard Model and Beyond

These insights lead to the Standard Model, see Fig. 1. It is fully renormalizable at the perturbative level, its infinities are fully under control, in particular because the anomalies cancel out, and it is very nearly asymptotically free, at least to such an extent that no corrections to the model are needed until extremely high energies would be reached, far beyond what is presently attainable in particle accelerators. The rest of this section will be used to admire the beauties of the Standard Model, without yet attempting to say anything new to the experts.

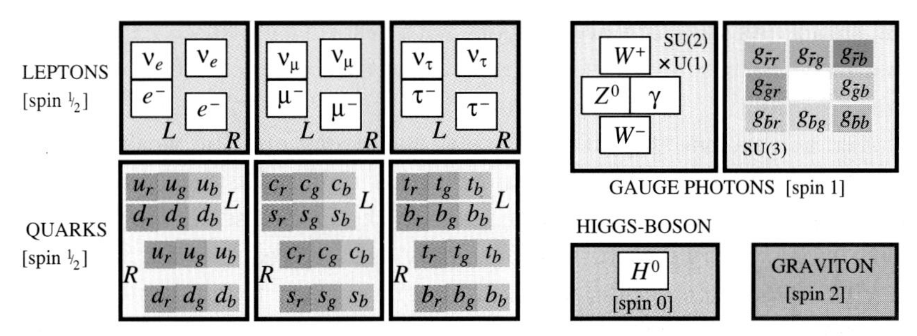

Fig. 1. The Standard Model of the Elementary Particles.

The Standard Model agrees surprisingly well with the experimental observations. Indeed, for more than 20 years, physicists have tried to detect deviations from the Standard Model predictions. What was found, instead, is slight improvements in the details that were known. Most of the objects that originally were still missing have been found — with as yet the only exception being the Higgs boson — most of the values of the freely adjustable physical constants were determined much more accurately, and some more delicate improvements were made: neutrinos received mass values and mixing angles much like their colored counterparts, the quarks.

The extraordinary successes of the Standard Model allowed us, for the first time, to speculate about Nature'a Laws at a domain of energies and distance scales far beyond what can be reached directly in most experiments. First, one may observe the great similarity between the quark quantum numbers and the lepton quantum numbers. The newly established fact that neutrinos have masses and mixing angles only further enhances this similarity. It is therefore natural to expect that $SU(3)_{\text{color}}$ and $U(1)$ emerge as unbroken subgroups from a spontaneously broken $SU(4)$. The scale at which this spontaneous breaking should occur will probably be extremely high, so as to allow for the strong coupling coefficient to run towards the $U(1)$ coupling strength. In any case, present precision measurements do not betray the presence of heavy vectors with lepto-quark quantum numbers.

Another extension of the Standard Model may be suggested by adding a right-handed $SU(2)$ gauge field, turning $SU(2)_{\text{left}} \times U(1)$ into $SU(2)_{\text{left}} \times SU(2)_{\text{right}}$. The algebra of this group coincides with that of $SO(4)$.

The algebra of the $SU(4)$ group of the strong interactions coincides with that of $SO(6)$. The last step is then to join $SO(4)$ and $SO(6)$ into $SO(10)$. The possibility of this last step is so delicate that it is seen as a strong argument in favor of the previous steps towards this construction. The fermions form three generations of a single spinorial **16** representation of this group. As they are also spinors with respect to ordinary space-time rotations, one may even suspect a Kaluza-Klein compactification process here.

Also at this last step in unification, the strong and the weak coupling strengths must converge. This is why this unification mechanism must happen at a unification scale which is near the Planck scale. However, the left-right symmetry restoration of the previous step requires the $SU(2)$ and the $U(1)$ forces to coincide, which also requires these to run until a similar unification scale is reached. The fact that these various steps of unification all appear to take place at nearly the same energy, close to the Planck scale, is an important piece of information to be drawn from our present understanding of the Standard Model.

The above arguments are of course not new. It is important however to stress that, so far, no supersymmetry or superstring theories were needed for these observations. Supersymmetry does allow us to construct a theory where the left-right symmetry braking and the $SO(10)$ breakdown happen *precisely* simultaneously. This is often seen as a clue pointing towards a supersymmetric extension of the Standard Model. This signal is perhaps not very strong, but it is also unsatisfactory to attribute it to coincidence.

3. Questions

While staring at these conspicuous and at the same time mysterious structures, we are prompted to ask many questions. Some of these questions have been discussed at numerous occasions by many authors. For instance:

- Why do there appear to be exactly three generations of quarks and leptons?
- Where do these wildly varying mass values come from? There appears to be a *hierarchy* of scales in Nature, generated by this hierarchy of mass values. Why do we have so many numbers whose orders of magnitude span such a wide range?
- Why in particular these groups and these representations? Is it a coincidence that the spinors also form a spinorial **16** representation of $SO(10)$? Why are the Higgs scalars arranged in the representations that we see?
- Is there supersymmetry?
- How do we include the gravitational force? Is it a coincidence that the Planck scale and the unification scale appear to be fairly close together?

These are the questions that we have been unable to answer, in spite of many vigorous attempts.

In this lecture, I would like to put more emphasis on questions that are not heard as often, while they may be equally important to ask:

- ◇ Can gravity (and, in particular, curved, closed universes) be reconciled with the rules of Quantum Mechanics at all? Can the Universe as a whole form states in a Hilbert space, while the concept of an "observer", who should be sitting outside this universe, would be a notion that is hardly acceptable?
- ◇ How important is the notion of *locality* in physics? Those who investigate the foundations of Quantum Mechanics, and those who try to interpret Super String Theory, are often tempted to drop locality as a necessary feature of the Laws of Physics. But can we make sense at all of a framework describing our world if we allow action-at-a-distance everywhere?
- ◇ How can *local gauge invariance* arise in a deeper fundamental theory?
- ◇ How can a *naturally* small (or vanishing) cosmological constant arise — while fermions and bosons only match approximately, if at all?

4. Holography

My motivation for asking these questions requires further clarification. We know that the gravitational force is fundamentally unstable. Unlike the electromagnetic case, like "charges" attract one another, so that gravity could become an accumulative force. In general relativity, the instability against implosion leads to one stationary solution in which all matter imploded completely: the black hole. There exist some misconceptions concerning black holes. Some authors appear to believe that the description of a black hole requires the assumption that matter forms a singularity first, and that, by assuming possible exotic laws of physics, one might be able to avoid the formation of a black hole altogether. However, the most characteristic feature of a black hole is the emergence of a *horizon*. The singularity could be affected by exotic laws of Nature, but this is largely immaterial for our understanding of a black hole. The emergence of the horizon is purely a consequence of General Relativity, requiring nothing more than the equivalence between inertial and gravitational mass, in combination with special relativity. Of course, one may question these principles of Nature, but one should realize that they have been corroborated by numerous experiments. The existence of black holes follows from laws of physics that all have been tested to great accuracy.

Applying directly the Laws of Quantum Field Theory to the environment of the horizon, S. Hawking[5] was led to conclude that particles of all species are emitted, and that they are distributed according to a black body spectrum with an easily computable temperature.[a] This beautiful result strongly suggests that black holes are much more mundane forms of matter than the esoteric pure "balls of gravity" that they appear to be in the classical theory of General Relativity. From their thermal behaviour, one can derive an estimate for the density of states, and from that, one can easily imagine that black holes will blend naturally into the spectrum of ordinary particles at the Planck scale. So, in a sense, black holes and particles are the same things, and, intuitively, we can hardly imagine otherwise.

The expression we obtain for the density of states corresponds to what one would get if all dynamical degrees of freedom of a black hole would be distributed evenly over the horizon. *There is one boolean degree of freedom on every* $7.24 \cdot 10^{-66}$ cm^2 *of the horizon.* This number is $4 \ln 2$ times the Planck length squared. It is important to realize that, in a linearized quantum field theory, the number of degrees of freedom would *diverge* on the horizon. The entropy of a black hole appears to match the degrees of freedom of a quantum field theory only if one performs a hard cut-off at distances comparable to the Planck length from the horizon.[7] At first sight, such a cut-off seems to be forbidden by General Relativity, but one must realize that, at that scale, gravitational forces become strong, so that linearized quantum field theory does not apply; exactly what to do is not properly understood. Applying to the theory with cut-off a general coordinate transformation back to locally regular space-time, one ends up with what is known as the "holographic principle": there cannot exist more physical degrees of freedom in any closed system than one per $7.24 \cdot 10^{-66}$ cm^2 of its surface area.[8]

It is here that the question of locality comes up: *How can one construct a theory which is reasonably consistent with locality features that we perceive in the real world and yet has its physical degrees of freedom distributed evenly over a surface?* Can such a theory be consistent with unitarity and causality?

A natural answer to this question may seem to be, that what is needed is a "topological" theory, a theory where all physical degrees of freedom

---

[a] Although the value of this temperature is agreed upon by most authors, one may suspect that it hinges on an assumption that is not yet proven to be correct. The author still maintains that an alternative scenario cannot be excluded.[6]

can be mapped onto the boundary. This is how holography is usually interpreted in string theory. But this does not appear to resolve the problem of locality. How can it be that, nevertheless, we perceive our world as $3+1$ dimensional? From a physical point of view, this seems to be a genuine *paradox*, not unlike the one that in 1900 lead Max Planck to postulate energy quantization. This time, however, one might be forced to arrive the opposite conclusion: the paradox came from our presumption that states form a quantum mechanical Hilbert space. There must be something wrong with that.[9]

At this point, I think it is totally legitimate to ask: "Why should these issues *not* be related to the question of the foundation of quantum mechanics?"

And so, one is lead back to the issue of (local?) hidden variables. It is an issue not to be forgotten.

5. Determinism

Rather than trying to devise tricky "Gedanken experiments" in which one could search for deterministic features, we start from the other end. Consider some simple *deterministic* system, consisting of a set of N states,

$$\{(0),(1),\ldots,(N-1)\}\,, \tag{5.1}$$

on a circle. Time is discrete, the unit time steps having length τ (the continuum limit is left for later). The evolution law is

$$t \to t+\tau \quad : \qquad (\nu) \to (\nu + 1 \operatorname{mod} N)\,. \tag{5.2}$$

Note that this model is representative for anything that shows periodic behaviour in time.

Introducing a basis for a Hilbert space spanned by the states (ν), the evolution operator can be written as

$$U(\Delta t = \tau) = e^{-iH\tau} = \begin{pmatrix} 0 & & & & 1 \\ 1 & 0 & & & \\ & 1 & 0 & & \\ & & \ddots & \ddots & \\ & & & 1 & 0 \end{pmatrix}. \tag{5.3}$$

The eigenstates of U are denoted as

$$|n\rangle = \frac{1}{\sqrt{N}} \sum_{\nu=1}^{N} e^{\frac{2\pi i n \nu}{N}} (\nu)\,, \qquad n = 0,\ldots,N-1\,. \tag{5.4}$$

This evolution law can be represented by a Hamiltonian using the notation of quantum physics:[b]

$$H|n\rangle = \frac{2\pi}{N_T}\, n|n\rangle\,. \tag{5.5}$$

This Hamiltonian can be used to describe a *quantum* harmonic oscillator. It is our proposal to identify the states (5.4) with the eigenstates of an harmonic oscillator. They evolve exactly as required. There is a slight problem, however: the number n is limited to be less than N, the number of discrete positions on the circle.

The remedy to this problem, at first sight, appears to be easy: we take the $N \to \infty$ limit. Indeed, then we get exactly all states of the harmonic oscillator.[10] However, upon further inspection, there still remains an important obstacle. If one wants to introduce interactions, that is, non-harmonic terms in the oscillator, expressions are needed that relate the "ontological states" (ν) to the harmonic oscillator states. One then discovers that an operator that links the ground state $|\nu = 0\rangle$ to the *highest* energy state $|\nu = N - 1\rangle$ (see Fig. 2(a)) continues to contribute to these relations in an essential way. Therefore, the continuum limit is not as smooth as what would be desired.

What we have presently under investigation is the question whether there exists a superior way to obtain a continuum limit. In the continuum limit, one obtains Fig. 2(b). If the evolution can run along the circle in *both* directions, the eigenvalues of the Hamiltonian will be plus or minus Eq. (5.5). We may be tempted to turn Fig. 2(b) into Fig. 2(c). Since there are then two classes of states, it appears that a fermionic degree of freedom arises. It is too early to conclude from this that supersymmetry has anything to do with a deterministic interpretation, but it could be that fermions will have a natural role to play in our deterministic models. Generally speaking, one finds that fermions are associated to discrete degrees of freedom in a deterministic underlying model. Unfortunately, there are still several problems with this picture.

Since it is easy to imagine that deterministic, periodic systems may influence one another, the inclusion of *interactions* should not give problems in principle. It seems, however, that the mathematical methods that are familiar to us do not apply. This holds in particular when one tries to incorporate information loss.

---

[b]If so desired, one may replace n by $n + \frac{1}{2}$, by adding a phase factor to the evolution operator U of Eq. (5.3).

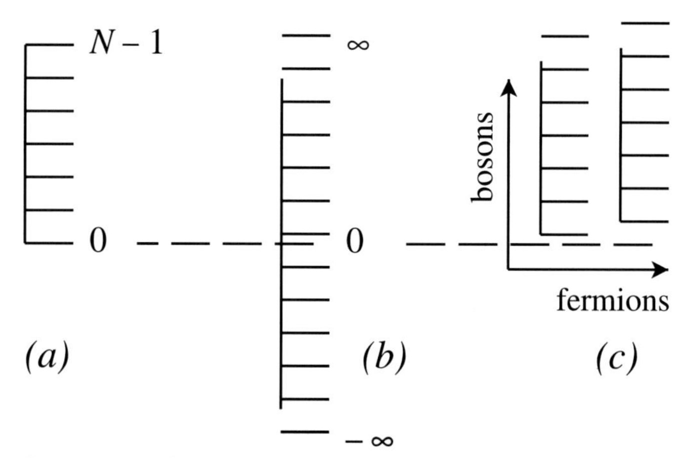

Fig. 2. The spectrum of the Hamiltonian for a deterministic, periodic system. (a) N points on a circle, rotating with one fixed angular velocity. (b) In the continuum limit, it is hard to ignore the negative energy solutions. (c) After a sign flip of the negative energy states, we get a double spectrum, suggesting a fermionic degree of freedom.

6. Information Loss

The reasons why *information loss* may be an essential ingredient in deterministic hidden variable models of the sort pioneered above, has been extensively discussed in Ref. 9. A prototype microcosmos with information loss is the model of Fig. 3. Following the arrows, one would conclude that the evolution matrix is

$$U = \begin{pmatrix} 0 & 0 & 1 & 0 \\ 1 & 0 & 0 & 1 \\ 0 & 1 & 0 & 0 \\ 0 & 0 & 0 & 0 \end{pmatrix} . \tag{6.1}$$

This, of course, is not a unitary matrix. One way to restore unitarity would be to remove state # 4. The problem with that is that, in universes with tremendously many allowed states, it would be very difficult to determine which of the states are like state number 4, that is, they have no state at all in their (distant) past.

A preferred way to proceed is therefore to introduce *equivalence classes* of states. Two states are equivalent iff, some time in the near future, they evolve into one and the same state.[c] In Fig. 3, states ## 1 and 4 are

[c]It could also happen that two states merge into the same state in the *distant* future, but in many models merging may become increasingly unlikely as time goes on.

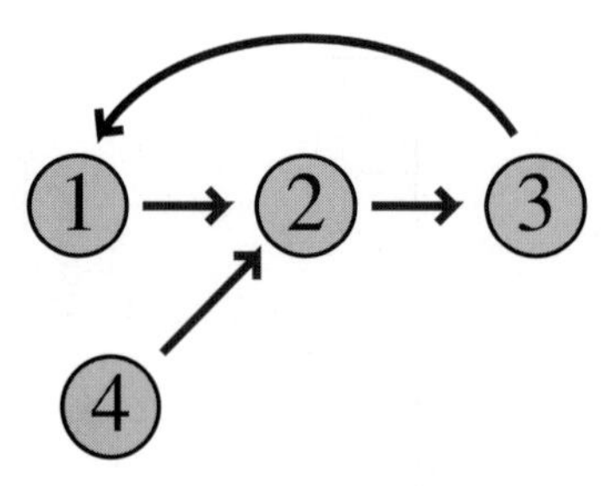

Fig. 3. Mini-universe with information loss. The arrows show the evolution law.

equivalent, so they form one class. By construction then, equivalence classes evolve uniquely into equivalence classes. In this respect, this universe is like the one described by Eq. (5.2) and the matrix U of Eq. (5.3), with $N = 3$.

It should be emphasized that, at the Planck scale, information loss is not a small effect but a very large effect. Large numbers of "ontological" states are in the same equivalence class, and the equivalence classes form a much smaller set than the class of all states. This is how it can happen that the total number of distinguishable quantum states (= the number of equivalence classes) may only grow exponentially with the *surface* of a system, whereas the total number of ontological states may rise exponentially with the volume. This seems to be demanded by black hole physics, when we confront the laws of quantum mechanics with those of black holes.

Information loss at the level of the underlying deterministic theory, may also explain the apparent lack of causality in the usual attempts to understand quantum mechanics in terms of hidden variables. The definition of an equivalence class refers to the future evolution of a system, and therefore it should not be surprising that in many hidden variable models, causality seems to be violated. One has to check how a system will evolve, which requires advance knowledge of the future.

Information loss at the Planck scale may also shed further light on the origin of gauge theories. it could be that, at the level of the ontological degrees of freedom at the Planck scale, there is no local gauge symmetry at all, but in order to describe a physical state, that is, an equivalence class, we need to describe a particular member of this class, a single state. Its relation to the other members of the same equivalence classes could be what is presently called a "gauge transformation".

7. Final Remarks

The fact that we need gauge-noninvariant vector potentials and other fields, to describe our world, is a peculiar fact of life that we learned from studying particle physics, not in the least due to the work of C.N. Yang. This, together with other very specific peculiarities of the Standard Model, is what was meant with the words "hidden information" in the title, although the title may also be regarded as referring to the apparent need for phenomena such as information loss in our basic theories.

An important remark about symmetries:[11] most of the symmetry groups of Nature refer to symmetries in Hilbert Space. In our pre-quantum theories, there will, presumably, be much less symmetry. We expect that the physical degrees of freedom become discrete at the Planck scale, and this implies that, at best, translation symmetry becomes discrete, whereas rotation symmetry may disappear altogether. If an ontological theory can be constructed (a theory describing a single reality), it is likely to require a new form of mathematics allowing us to introduce symmetries relating *ontological operators* (which we call "beables") to *quantum operators* (operators that replace an ontological state by an other state, called changeables). it is these symmetries that then will lead to the symmetries we know from particle physics.

References

1. C.N. Yang and R.L. Mills, *Phys. Rev.* **96** (1954) 191.
2. H.B. Newman and Th. Ypsilantis, *History of Original Ideas and Basic Discoveries in Particle Physics*, Erice, July/Aug. 1994, Plenum Press, New York and London, ISBN 0-306-45217-0.
3. F. Englert and R. Brout, *Phys. Rev. Lett.* **13** (1964) 321.
4. P.W. Higgs, *Phys. Lett.* **12** (1964) 132; *Phys. Rev. Lett.* **13** (1964) 508; *Phys. Rev.* **145** (1966) 1156.
5. S.W. Hawking, *Commun. Math. Phys.* **43** (1975) 199; J.B. Hartle and S.W. Hawking, *Phys. Rev.* **D13** (1976) 2188.
6. G. 't Hooft, *J. Geom. Phys.* **1** (1984) 45.
7. G. 't Hooft, *Nucl. Phys.* **B256** (1985) 727.
8. G. 't Hooft, *Dimensional Reduction in Quantum Gravity*, Essay dedicated to Abdus Salam, Utrecht preprint THU-93/26; `gr-qc/9310026`; *Black Holes and the Dimensionality of Space-Time.* Proceedings of the Symposium "The Oskar Klein Centenary", 19-21 Sept. 1994, Stockholm, Sweden. Ed. U. Lindström, World Scientific 1995, 122-137.
9. G. 't Hooft, *Class. Quant. Grav.* **16** (1999) 3263; also published in *Fundamental Interactions: From Symmetries to Black Holes* (conference held on the occasion of the "Emeritat" of Franois Englert, 24-27 March 1999, ed. by J.-M. Frere et al., Univ. Libre de Bruxelles, Belgium, p. 221, `gr-qc/9903084`.

10. G. 't Hooft, *Determinism in Free Bosons*, SPIN-2001/07, ITP-UU-01/14, `hep-th /0104080`; id., *Quantum Mechanics and Determinism*, in Proceedings of the Eighth Int. Conf. on Particles, Strings and Cosmology, Univ. of North Carolina, Chapel Hill, Apr. 10-15, 2001, P. Frampton and J. Ng, Eds., Rinton Press, Princeton, pp. 275-285; ITP-UU/01/18, SPIN-2001/11, `hep-th/0105105`.
11. G. 't Hooft, *Determinism Beneath Quantum Mechanics*, Proceedings Temple University Conf. "Quo vadis Quantum Mechanics?", Sept. 2002, ITP-UU-02/69, SPIN-2002/45.

Maurice Goldhaber

THE $\theta-\tau$ PUZZLE REVISITED

DEDICATED TO PROFESSOR C. N. YANG

MAURICE GOLDHABER

Brookhaven National Laboratory, Upton, NY 11973, USA

October 1993

After Dalitz[1] deduced from a detailed analysis of experimental data that the decay products of the apparently "like" mesons θ and τ, into 2π and 3π, respectively, form states of opposite parity, attempts were made to find differences in some of the properties of θ and τ *before* decay, by measuring the mass, lifetime, and scattering for each. Within the accuracy attained no differences were found, but from such experiments a definitive proof of identity cannot be reached, even with "asymptotically" improved acuracy. Since identity would have implied the non-conservation of parity, contradicting an important symmetry principle, thus the θ–τ puzzle was born!

Soon after the April 1956 Rochester Conference,[2] where this puzzle was much discussed, a brilliant insight by Lee and Yang[3] showed the way to a solution. They cut the Gordian knot by asking a more general question: Is parity conserved in weak interactions? Nature answered no, first for β and μ decay,[4] and later for other decays by weak interactions. After the non-conservation of parity was established, it was taken for granted that θ and τ were identical, and their decays were considered to be just two of the several branches of kaon decay.

Because a statement that two "like" particles are identical is an absolute one, it can be proved only if we use well-established fundamental principles. The identitiy of particles, either bosons or fermions, may be shown explicitly if their wavefunction is found to be either symmetric (Bose–Einstein statistics) or antisymmetric (Fermi–Dirac statistics). But "like" particles of finite lifetime have a finite mass distribution, and while at first sight, they may thus not appear to be identical, the uncertainty relation tells us that during the finite time of the experiment a difference in their masses cannot be discerned; therefore, the particles are still *indistinguishable*, which is both necessary and sufficient for a wavefunction of well-defined symmetry. In other words, it is sufficient for "like" particles to have an identical *complex* mass to be indistinguishable.[5]

With the help of the Pauli exclusion principle, a symmetry argument was used to establish the identity of β rays with atomic electrons, often still investigated by an "asymptotic" approach about half a century ago, until their identity was demonstrated by showing that β-rays at the end of their path cannot fall into an already full atomic electron shell.[6]

For the θ and τ bosons it may be possible to demonstrate the identity by looking for a correlation between those pairs of kaons (produced in particle or nuclear interactions) that finally decay, *one* into a θ and the *other* into a τ. The techniques of studying Bose–Einstein correlations were developed for pairs of photons,[7] pions[8] and kaons,[9] and are used to assess the size of their originating sources. For pions and kaons the techniques only came along *after* Lee and Yang's paper. The correlations explored between kaons, copiously produced in heavy-ion collisions, disregarded their final decay modes.[10] Quantum mechanics tells us that a particle does not "know" the timing and manner of its decay, but it may be of interest to test the independence of both the decay times and decay modes of *entangled* particles when they decay in flight. They would lose coherence when they are stopped.

References

1. R. H. Dalitz, *Phil. Mag.* **44**, 1068 (1953).
2. J. Ballam, V. L. Fitch, J. Fulton, K. Huang, R. R. Rau and S. B. Treimen (eds.), *Proc. Sixth Annual Rochester Conf.* (Interscience Publishers, NY, 1956).
3. T. D. Lee and C. N. Yang, *Phys. Rev.* **104**, 254 (1956).
4. C. S. Wu, E. Ambler, R. W. Hayward, D. D. Hoppes and R. Hudson, *Phys. Rev.* **105**, 1413 (1957); J. I. Friedman and V. L. Telegdi, **105**, 1682 (1957); R. Garwin, L. Lederman and M. Weinrich, *ibid.* **105**, 14151 (1957).
5. M. Goldhaber and J. Weneser, *Festschrift in Honor of Yuval Ne'eman. From SU(3) to Gravity*, eds. E. Gotsman and G. Tauber (Cambridge University Press, 1985), p. 107.
6. M. Goldhaber and G. Scharff-Goldhaber, *Phys. Rev.* **73**, 1472 (1948).
7. R. Hanbury-Brown and R. Q. Twiss, *Phil. Mag.* **45**, 663 (1954).
8. G. Goldhaber, S. Goldhaber, W. Lee and A. Pais, *Phys. Rev.* **120**, 300 (1960).
9. T. Akkeson *et al.*, *Phys. Lett.* **B155**, 128 (1985).
10. Y. Akiba *et al.*, *Phys. Rev. Lett.* **70**, 1057 (1993).

ITP Presentation. *From left:* Roger Behrend, Hwa Tung Nieh, Werner Vogelsang, John Smith, Gary Shiu, William Weisberger, Peter van Nieuwenhuizen, George Sterman, Robert Shrock, Vladimir Korepin, Gerald E. Brown, Martin Rocek, Ernest D. Courant, Warren Siegel and Alfred Goldhaber.

REMARKS FOR ITP PRESENTATION, C. N. YANG SYMPOSIUM

GEORGE STERMAN*

C. N. Yang Institute for Theoretical Physics, Stony Brook University, Stony Brook, NY 11794-3840, USA

22 May 1999

Tonight we celebrate one milestone in a distinguished career, and also an ongoing constellation of achievements that have touched, in one way or another, every person in this room.

Frank, as members of the Institute for Theoretical Physics of the University at Stony Brook, our most appropriate expression of thanks surely will be to continue in the best tradition of research in physics. Still, we felt the need to mark the occasion of your retirement as Director by presenting you with something tangible, as a token of our regard.

We have your own words as testimony of the historical significance of a certain article in the journal *Zeitschrift für Physik*, from 1922. This article, by Erwin Schrödinger, is entitled (in translation) "On a remarkable property of the quantum orbit of one electron". In it, Schrödinger noticed that a certain expression — a gauge factor — that depends on the position of an electron moving in an atomic orbit returns to unity every time the electron goes around — *if* one introduces into it the number i, the square root of minus one. With this simple substitution, Schrödinger opened the door to the interference phenomena that are the hallmark of quantum mechanics. You have argued, in a volume from the 1987 centenary of Schrödinger's birth, that it was this observation that eventually led to the Schrödinger equation and, through work of Hermann Weyl and Fritz London, to the description of fundamental interactions in terms of gauge fields.

Like a milestone in a career, the article of Schrödinger stands as a turning point in the history of science. Its publication appears now as one "still point of the turning world", in the words of a poem by T. S. Eliot that you quoted at your seventieth birthday celebrations a few years ago.

Eventually, a book is shelved away, but great ideas have another kind of life. In this article of Schrödinger's we discover the origin, perhaps one of many, of a

*For the members of the Institute for Theoretical Physics.

profound insight. The documents that recall that time remain as symbols of the continuity of thought, and it is as such that we offer this original bound copy of Volumes 11 and 12 of *Zeitschrift für Physik*, 1922, which is graced as well with articles by Born, Einstein, Ehrenfest and other luminaries of the period.

As a text and as an object, it provides a view of an important moment from one of the many scientific stories in which you later played a central role: the development of gauge symmetries. This story extends from Faraday and Maxwell in the nineteenth century, through the discovery of quantum mechanics nearly eighty years ago, to the present and into the future, with the Yang–Mills paper the beginning of the modern chapter. Gauge theories have become the language of all the fundamental interactions, in the Standard Model and beyond.

We present this volume with our thanks as members of the Institute for Theoretical Physics, for defining the Institute with your more than thirty years (so far) of active presence, for your leadership by example.

We also give it with the sense that we are indebted for the insights you have shared, in publication and in conversation. There are many such classics, which some of us first saw when they were new, and some of us first learned from textbooks. We are looking forward to yet others.

But it goes beyond even that. The beautiful discoveries you have made, and the elegant methods you have created, continue to enable us, in our best moments, to make discoveries and creations of our own, on which we can look back with pride.

In this spirit, we present you with this little historic document, with our congratulations, our thanks, and our best wishes.

Although your duties as Director are behind you, we are proud to count you as a colleague still.

References

1. E. Schrödinger, *Zeitschrift für Physik* **12** (1922–1923) 13.
2. C. N. Yang, in *Schrödinger: Centenary Celebration of a Polymath*, ed. C. W. Kilmister, Cambridge University Press, 1987, p. 53.

Über eine bemerkenswerte Eigenschaft der Quantenbahnen eines einzelnen Elektrons.

Von **Erwin Schrödinger** in Zürich.

(Eingegangen am 5. Oktober 1922.)

In der Weylschen Weltgeometrie[1] tritt außer der bekannten quadratischen Form der Differentiale, welche die Metrik in den einzelnen Weltpunkten bestimmt, noch eine Linearform auf

$$\varphi_0 dx_0 + \varphi_1 dx_1 + \varphi_2 dx_2 + \varphi_3 dx_3 = \varphi_i dx_i,$$

welche den metrischen Zusammenhang der Weltpunkte untereinander festlegt. Ihre geometrische Bedeutung ist die, daß die Maßzahl einer „Strecke“ l (Quadrat des Absolutbetrages eines Vektors) bei „kongruenter Verpflanzung“ der Strecke in einen Nachbarpunkt nicht ungeändert bleibt, sondern die Änderung

$$dl = -l\varphi_i dx_i \tag{1}$$

erleidet. Weyl hat entdeckt, daß durch beides zusammen (Metrik der einzelnen Weltpunkte + metrischen Zusammenhang) ein affiner Zusammenhang der Welt (d. h. der Begriff der Parallelverschiebung eines Vektors) mitgegeben ist, wenn man nur verlangt, daß bei Parallelverschiebung eines Vektors seine Strecke auch kongruent verpflanzt werden soll. Bei kongruenter Verpflanzung einer Strecke entlang eines endlichen Stückes einer Weltlinie — z. B. bei Parallelverschiebung eines Vektors entlang eines solchen Stückes — multipliziert sich die Maßzahl der Strecke mit dem Faktor

$$e^{-\int \varphi_i dx_i}, \tag{2}$$

wobei das Linienintegral natürlich dem betreffenden Weltlinienstück entlang zu nehmen ist und wesentlich vom Wege abhängt, wofern die Größen

$$f_{ik} = \frac{\partial \varphi_i}{\partial x_k} - \frac{\partial \varphi_k}{\partial x_i} \tag{3}$$

nicht identisch verschwinden. — Physikalisch bilden die Komponenten des oben erwähnten affinen Zusammenhangs das Gravitationsfeld, die f_{ik} das elektromagnetische Feld. Liegen die Verhältnisse so — und ist die Koordinatenwahl so getroffen —, daß in einem Weltgebiet wenigstens mit einer gewissen Annäherung x_0 die Zeit (in sec) und $x_1 x_2 x_3$ kartesische Koordinaten (in cm) sind, so sind die φ_i bis auf

---

[1]) Siehe etwa H. Weyl, Raum, Zeit, Materie, 4. Aufl. Berlin, Springer, 1921. — Unten mit Weyl, RZM zitiert.

SCIENCES

PROGRAM

Friday, May 21

R. Mills, Chair

9.00 – 9.30	J. Zinn-Justin	Renormalization of gauge theories and master equation
9.30 – 10.00	W. Bardeen	Static quantities in the standard model
10.00 – 10.30	S. C. Zhang	High T_c superconductivity: Symmetries and reflections

G. Chew, Chair

11.00 – 11.30	R. Baxter	Solvable models in statistical mechanics: Ising to chiral Potts
11.30 – 12.00	L. Cooper	Memories and memory
12.00 – 12.30	W. Ketterle	Dilute Bose–Einstein condensates — early predictions and recent experimental studies

C. Jarlskog, Chair

2.00 – 2.30	G. Brown	Yang–Mills theory and chiral restoration in nuclear physics
2.30 – 3.00	M. Veltman	Reconsideration of a question posed by Frank Yang at the Bonn Conference in 1973
3.00 – 3.30	J. Cronin	Experimental program in CP violation — 35 years

J. Simons, Chair

4.00 – 4.30	L. Alvarez-Gaume	Some remarks on recent measurements of T-reversal violation
4.30 – 5.00	I. Singer	Line bundles on 6 manifolds
5.00 – 5.30	S. T. Yau	Some application of ideas of modern physics to geometry

Saturday, May 22

P. Chu, Chair

9.00 – 9.30	B. Sutherland	Bethe's Ansatz: Now and then
9.30 – 10.00	A. Tonomura	The microscopic world unveiled by electron waves
10.00 – 10.30	L. Faddeev	Wu–Yang monopole revisited

E. Ambler, Chair

11.00 – 11.30	E. Courant	Possibilities of spin physics at high energy
11.30 – 12.00	M. Rosenbluth	The quest for fusion energy
12.00 – 12.30	L. H. Yu	Research and development towards x-ray free electron laser

G. Zhou, Chair

2.00 – 2.30	T. T. Wu	Some remarks on Yang–Mills theory
2.30 – 3.00	G 't Hooft	Holography and a deterministic interpretation of quantum mechanics
3.00 – 3.30	S. Ting	The AMS experiment

V. Fitch, Chair

4.00 – 4.30	J. Steinberger	Some bits of neutrino history
4.30 – 5.00	M. Goldhaber	Neutrino oscillations

Photographs taken at the Symposium

Facing page: C. N. Yang with Robert L. Mills (upper left), Rodney Baxter (upper right), John Wheeler (lower left), "Chicago classmates" (lower right): Lincoln Wolfenstein, Geoffrey Chew, Marshall Rosenbluth and Jack Steinberger (from left)

Next page: Ernest D. Courant and Freeman Dyson (upper left); C. N. Yang with the first ITP postdocs (lower left): Michael Nieto, Hwa-Tung Nieh, William Bardeen, Wu-Ki Tung (from left); C. N. Yang with Peter van Nieuwenhuizen and Martinus Veltman (lower right)

Following page: C. N. Yang and Chi Li Yang with celebrants, with Alexander Chao (lower right).

STONY

SCIENCES

PARTICIPANTS

Allen, Philip B.
Alvarez-Gaume, Luis
Ambler, Ernest
An, Jianji (companion to Zhou)
Bardeen, Bill
Bastianelli, Fiorenzo
Baxter, Rodney
Behrend, Roger
Benka, Stephen
Bergeman, Tom
Bigeleisen, Jacob
Boer, Daniel
Brown, Gerald E.
Castellani, Leonardo
Chalmers, Gordon
Chan, F. T.
Chang, Lay Nam
Chang, Ngee-Pong
Chang, Sherry
Chao, Alexander
Chau, Ling-Lie
Chen, W.
Chen, Wei
Cheng, Hai-Yang
Cheng, Hung
Cheng, Kuo-Shung
Chew, Geoffrey
Chiang, Tsai-Chien
Cho, Y. M.
Christ, Tassilo
Chu, Benjamin
Chu, Paul C. W.
Clifford, Swartz
Cooper, Leon
Courant, Ernest D.
Cronin, James
Cronin, Jim
de Wit, Bernard
Deng, Yuefan
DeWitt, Bryce
DeWitt-Morette, Cecile
DeZafra, Robert
Faddeev, Ludwig
Fan, Chao-hui
Fan, Chungpeng
Fan, Shih-Fang
Feingold, Arnold
Fernandez-Labastida, Jose
Finocchiaro, Guido
Fischbach, Ephraim
Fischler, Mark
Fitch, Val
Freedman, Dan
Fujikawa, Kazuo
Goldhaber, Alfred
Goldhaber, Maurice
Grannis, Paul
Groeneveld, Hans
Guo, Xiaofeng
Ha, Yuan K.
Hanes, Carl
Hasslacher, Brosl
Hayot, Fernand
Hill, C. Denson
Horwitz, Larry
Hoyer, Paul

Hwang, W.-Y. Pauchy
Itoyama, Hiroshi
Jain, Jainendra
Jarlskog, Cecilia
Jung, Chang Kee
Kahn, Peter
Ketterle, Wolfgang
Kilgore, William
Kinoshita, Toichiro
Kirz, Janos
Kong, Xiangpeng
Korepin, Vladimir
Kuo, Thomas T. S.
Lam, Harry
Lee, Linwood
Lee, T. K.
Leung, George
Li, Bing An
Lindstrom, Ulf
Liu, Keh-fei
Lourie, Robert
Low, Francis
Lu, Wentao
Lukens, James
Mar, Gary
Marciano, William
Marx, Michael
McCarthy, Robert L.
Meng, Ta-chung
Mills, Robert L.
Milnor, John
Ng, Kim K.
Ng, Jack
Nieh, H. T.
Nielsen, Niels Kjar
Nieto, Michael Martin
O'Keeffe, Kami
Osland, Per
Park, Kwon
Paul, Peter
Peiliker, Patricia
Perk, Jacques H. H.
Phillips, Anthony
Pierce, Damien
Pisarski, Rob
Pond, T. Alexander
Prasad, Manoj
Qin, Jing
Qiu, Jianwei
Rafferty, Maureen A.
Rijssenbeek, Michael
Rocek, Martin
Rollet, Genevieve
Rosenbluth, M.
Samios, N. P.
Sasaki, Shoichi
Shamash, Yacov
Shiu, Gary
Shizuya, Ken-ichi
Shrock, Robert
Siegel, Diane
Siegel, Warren
Singer, I.
Sirlin, Alberto
Smith, Jack
Sprouse, Gene
Steinberger, Jack
Stell, George
Sterman, George
Sucher, Joe
Sullivan, Dennis
Sutherland, Bill
Swartz, Jerome
't Hooft, Gerard
Takhtajan, Leon
Thomas, Kisha
Ting, S.
Tonomura, Akira
Treiman, Sam
Tsai, Shan-Ho
Tsao, Hung-sheng
Tu, Eva

Tung, Michael
Tung, Wu-ki
Turpin, Catherine
van Baal, Pierre
van Nieuwenhuizen, Peter
Veltman, M.
Vepstas, Linas
Vogelsang, Werner
Wali, Kameshwar C.
Wang, Jin
Wei, Jie
Weisberger, William
Weng, Wu-Tsung
Wheeler, John A.
Wightman, Arthur
Wijers, Ralph
Wirzba, Andreas
Wolfenstein, Lincoln
Wu, A. C. T.
Wu, T. T.
Wu, Tai Tsun
Wu, F. Y.
Wu, Sau Lan
Wu, Yong-Shi
Xian, Ding Chang
Xu, Nu
Xu, Rui-Ming
Yang, C. P.
Yang, Chen-Yu
Yao, York-Peng E.
Yasui, Yoshiaki
Yau, S.-T.
Young, K. C.
Yu, Li Hua
Yu, Hoi-Lai
Zanelli, Jorge
Zee, A.
Zhang, Shoucheng
Zhang, Minghua
Zhou, Guang-Zhao
Zinn-Justin, Jean

Scientific and Related Works of Chen Ning Yang

[42a] C. N. Yang.
Group Theory and the Vibration of Polyatomic Molecules.
B.Sc. thesis, National Southwest Associated University (1942).

[44a] C. N. Yang.
On the Uniqueness of Young's Differentials.
Bull. Amer. Math. Soc. 50, 373 (1944).

[44b] C. N. Yang.
Variation of Interaction Energy with Change of Lattice Constants and Change of Degree of Order.
Chinese J. of Phys. 5, 138 (1944).

[44c] C. N. Yang.
Investigations in the Statistical Theory of Superlattices.
M.Sc. thesis, National Tsing Hua University (1944).

[45a] C. N. Yang.
A Generalization of the Quasi-Chemical Method in the Statistical Theory of Superlattices.
J. Chem. Phys. 13, 66 (1945).

[45b] C. N. Yang.
The Critical Temperature and Discontinuity of Specific Heat of a Superlattice.
Chinese J. Phys. 6, 59 (1945).

[46a] James Alexander, Geoffrey Chew, Walter Salove, Chen Yang.
Translation of the 1933 Pauli article in Handbuch der Physik, Vol. 14, Part II; Chapter 2, Section B.

[47a] C. N. Yang.
On Quantized Space-Time.
Phys. Rev. 72, 874 (1947).

[47b] C. N. Yang and Y. Y. Li.
General Theory of the Quasi-Chemical Method in the Statistical Theory of Superlattices.
Chinese J. Phys. 7, 59 (1947).

[48a] C. N. Yang.
On the Angular Distribution in Nuclear Reactions and Coincidence Measurements.
Phys. Rev. 74, 764 (1948).

[48b] S. K. Allison, H. V. Argo, W. R. Arnold, L. del Rosario, H. A. Wilcox, and C. N. Yang.
Measurement of Short Range Nuclear Recoils from Disintegrations of the Light Elements.
Phys. Rev. 74, 1233 (1948).

[48c] C. N. Yang.
On the Angular Distribution in Nuclear Reactions and Coincidence Measurements.
Ph.D. thesis, University of Chicago (1948).

[49a] T. D. Lee, M. Rosenbluth, and C. N. Yang.
Interaction of Mesons with Nucleons and Light Particles.
Phys. Rev. 75, 905 (1949).

[49b] E. Fermi and C. N. Yang.
Are Mesons Elementary Particles?
Phys. Rev. 76, 1739 (1949).

[50a] C. N. Yang.
Selection Rules for the Dematerialization of a Particle into Two Photons.
Phys. Rev. 77, 242 (1950).

[50b] C. N. Yang.
Possible Experimental Determination of Whether the Neutral Meson is Scalar or Pseudoscalar.
Phys. Rev. 77, 722 (1950).

[50c] C. N. Yang and J.Tiomno.
Reflection Properties of Spin 1/2 Fields and a Universal Fermi-Type Interaction.
Phys. Rev. 79, 495 (1950).

[50d] C. N. Yang and David Feldman.
The S-Matrix in the Heisenberg Representation.
Phys. Rev. 79, 972 (1950).

[51a] Geoffrey F. Chew, M. L. Goldberger, J. M. Steinberger, and C. N. Yang.
A Theoretical Analysis of the Process $\pi^+ + dp + p$.
Phys. Rev. 84, 581 (1951).

[51b] C. N. Yang.
Actual Path Length of Electrons in Foils.
Phys. Rev. 84, 599 (1951).

[52a] C. N. Yang.
The Spontaneous Magnetization of a Two-Dimensional Ising Model.
Phys. Rev. 85, 808 (1952).

[52b] C. N. Yang and T. D. Lee.
Statistical Theory of Equations of State and Phase Transitions. I. Theory of Condensation.
Phys. Rev. 87, 404 (1952).

[52c] T. D. Lee and C. N. Yang.
Statistical Theory of Equations of State and Phase Transitions. II. Lattice Gas and Ising Model.
Phys. Rev. 87, 410 (1952).

[52d] C. N. Yang.
Letter to E. Fermi dated May 5, 1952.
Unpublished.

[52e] C. N. Yang.
Special Problems of Statistical Mechanics, Part I and II.
Lectures given at University of Washington, Seattle,
April-July 1952. Notes taken by F. J. Blatt and R. L. Cooper.
Mimeographed and distributed by University of Washington.

[53a] C. N. Yang.
Report on Cosmotron Experiments.
Proc. International Conference on Theoretical Physics, Tokyo:
Science Council of Japan, 1954, p. 137.

[53b] Chen Ning Yang.
Recent Experimental Results at Brookhaven.
Proc. International Conference on Theoretical Physics, p.170
Science Council of Japan, 1954, (Tokyo).

[53c] C. N. Yang and R. Christian
Meson Spectrum from Cosmotron Target.
Brookhaven Internal Report, December 29, 1953, unpublished.

[54a] G. A. Snow, R. M. Sternheimer and C. N. Yang.
Polarization of Nucleons Elastically Scattered from Nuclei.
Phys. Rev. 94,1073 (1954).

[54b] C. N. Yang and R. Mills.
Isotopic Spin Conservation and a Generalized Gauge Invariance.
Phys. Rev. 95, 631 (1954).

[54c] C. N. Yang and R. L. Mills.
Conservation of Isotopic Spin and Isotopic Gauge Invariance.
Phys. Rev. 96, 191 (1954).

[54d] Chen Ning Yang.
Introduction to High Energy Physics.
Lecture given in summer of 1954 at Ann Arbor, Michigan.
Incomplete and unpublished.

[55a] C. N. Yang.
Talk at 1955 Rochester Conference, Session on High Energy Pion Phenomena.
High Energy Nuclear Physics, 1955. New York: Wiley Interscience Publishers, pp. 37-38.

[55b] T. D. Lee and C. N. Yang.
Conservation of Heavy Particles and Generalized Gauge Transformations.
Phys. Rev. 98, 1501 (1955).

[56a] K. M. Case, Robert Karplus, and C. N. Yang.
Strange Particles and the Conservation of Isotopic Spin.
Phys. Rev. 101, 874 (1956).

[56b] K. M. Case, Robert Karplus, and C. N. Yang.
Experiments with Slow K Mesons in Deuterium and Hydrogen.
Phys. Rev. 101, 358 (1956).

[56c] T. D. Lee and C. N. Yang.
Mass Degeneracy of the Heavy Mesons.
Phys. Rev. 102, 290 (1956).

[56d] T. D. Lee and C. N. Yang.
Charge Conjugation, a New Quantum Number G, and Selection Rules Concerning a Nucleon-Antinucleon System.
Il Nuovo Cimento, 10 (3), 749 (1956).

[56e] C. N. Yang.
Introductory Talk at the 1956 Rochester Conference, Session on Theoretical Interpretation of New Particles.
High-Energy Nuclear Physics, 1956, New York: Wiley Interscience Publishers.

[56f] C. N. Yang.
Expanding Universes by E. Schrödinger.
Science 124, 370 (1956).

[56g] Kerson Huang and C. N. Yang.
Quantum Mechanical Many-Body Hard Core Interactions.
Bull. Amer. Phys. Soc. 2 (1), 222 (1956).

[56h] T. D. Lee and C. N. Yang.
Question of Parity Conservation in Weak Interactions.
Phys. Rev. 104, 254 (1956).

[56i] T. D. Lee and C. N. Yang.
Possible Interference Phenomena Between Parity Doublets.
Phys. Rev. 104, 822 (1956).

[57a] K. Huang and C. N. Yang.
Quantum Mechanical Many-Body Problem with Hard Sphere Interaction.
Phys. Rev. 105, 767 (1957).

[57b] K. Huang, C. N. Yang, and J. M. Luttinger.
Imperfect Bose Gas with Hard Sphere Interaction.
Phys. Rev. 105, 776 (1957).

[57c] K. M. Case, R. Karplus, and C. N. Yang.
A Reply to a Criticism by Mr. A. Gamba.
Il Nuovo Cimento 5, 1004 (1957).

[57d] C. N. Yang.
Present Knowledge about the New Particles. Lecture given at the Seattle International Conference on Theoretical Physics, Sept. 1956.
Rev. Mod. Phys. 29, 231 (1957).

[57e] T. D. Lee, Reinhard Oehme, and C. N. Yang.
Remarks on Possible Noninvariance Under Time Reversal and Charge Conjugation.
Phys. Rev. 106, 340 (1957).

[57f] T. D. Lee and C. N. Yang.
Parity Nonconservation and a Two-Component Theory of the Neutrino.
Phys. Rev. 105, 1671 (1957).

[57g] T. D. Lee and C. N. Yang.
Derivative Coupling for μ Meson Decay in a Two-Component Theory of the Neutrino.
Unpublished.

[57h] T. D. Lee and C. N. Yang.
Many-Body Problem in Quantum Mechanics and Quantum Statistical Mechanics.
Phys. Rev. 105, 1119 (1957).

[57i] T. D. Lee, Kerson Huang, and C. N. Yang.
Eigenvalues and Eigenfunctions of a Bose System of Hard Spheres and Its Low-Temperature Properties.
Phys. Rev. 106, 1135 (1957).

[57j] T. D. Lee, J. Steinberger, G. Feinberg, P. K. Kabir, and C. N. Yang.
Possible Detection of Parity Nonconservation in Hyperon Decay.
Phys. Rev. 106, 1367 (1957).

[57k] Chen Ning Yang.
Lois de Symetrie et Particules Etranges.
Lecture given at Univ. Paris, May 1957. Lecture notes taken by Froissard and Mandelbrojt.
Unpublished.

[57l] T. D. Lee and C. N. Yang.
Errata: Question of Parity Conservation in Weak Interactions.
Phys. Rev. 106, 1371 (1957).

[57m] Kerson Huang, C. N. Yang, and T. D. Lee.
Capture of μ^- Mesons by Protons.
Phys. Rev. 108, 1340 (1957).

[57n] T. D. Lee and C. N. Yang.
Possible Nonlocal Effects in μ Decay.
Phys. Rev. 108, 1611 (1957).

[57o] T. D. Lee and C. N. Yang.
General Partial Wave Analysis of the Decay of a Hyperon of Spin 1/2.
Phys. Rev. 108, 1645 (1957).

[57p] T. D. Lee and C. N. Yang.
Elementary Particles and Weak Interactions.
BNL 443 (T-91) BNL (1957).

[57q] Kerson Huang, T. D. Lee, and C. N. Yang.
Quantum Mechanical Many-Body Problem and the Low Temperature Properties of a Bose System of Hard Spheres.
Lecture given at the Stevens Conference on the Many-Body Problem, January 1957. The Many-Body Problem, ed. by J. K. Percus, New York: Wiley Interscience 1963, p.165.

[57r] Chen Ning Yang.
Le Probleme a Plusieurs Corps en Mecanique Quantique et en Mecanique Statistique.
Lecture given at University of Paris June 1957. Lecture notes taken by C. Bouchiat and A. Martin. Unpublished.

[57s] C. N. Yang.
The Law of Parity Conservation and Other Symmetry Laws of Physics.
Les Prix Nobel. Stockholm: The Nobel Foundation (1957), p. 95. Also Science 127, 565 (1958).

[57t] C. N. Yang.
Nobel Banquet Speech, December 10, 1957.
Les Prix Nobel. Stockholm: The Nobel Foundation,1957, p. 53.

在諾貝爾賀宴上的講話
《楊振寧談科學發展》, 張美曼編,
八方文化企業公司(1992), p.15

[57u] C. N. Yang.
Short Autobiography.
Les Prix Nobel en 1957.

[58a] T. D. Lee and C. N. Yang.
Possible Determination of the Spin of Λ^0 from its Large Decay Angular Asymmetry.
Phys. Rev. 109, 1755 (1958).

[58b] J. Bernstein, T. D. Lee, C. N. Yang, and H. Primakoff.
Effect of the Hyperfine Splitting of a μ^- Mesonic Atom on its Lifetime.
Phys. Rev. 111, 313 (1958).

[58c] M. Goldhaber, T. D. Lee, and C. N. Yang.
Decay Modes of a $(\theta + \theta)$ System.
Phys. Rev. 112, 1796 (1958).

[58d] T. D. Lee and C. N. Yang.
Low-Temperature Behavior of a Dilute Bose System of Hard Spheres. I. Equilibrium Properties.
Phys. Rev. 112, 1419 (1958).

[59a] T. D. Lee and C. N. Yang.
Low -Temperature Behavior of a Dilute Bose System of Hard Spheres II. Nonequilibrium Properties.
Phys. Rev. 113, 1406 (1959).

[59b] T. D. Lee and C. N. Yang.
Many-Body Problem in Quantum Statistical Mechanics. I. General Formulation.
Phys. Rev. 113, 1165 (1959).

[59c] C. N. Yang.
Symmetry Principles in Modern Physics.
Lecture given at the 75th Anniversary Celebration of Bryn Mawr College, Session on Symmetries, November 6, 1959. Unpublished.

[59d] T. D. Lee and C. N. Yang.
Many-Body Problem in Quantum Statistical Mechanics. II. Virial Expansion for Hard-Sphere Gas.
Phys. Rev. 116, 25 (1959).

[60a] T. D. Lee and C. N. Yang.
Many-Body Problem in Quantum Statistical Mechanics. III. Zero-Temperature Limit for Dilute Hard Spheres.
Phys. Rev. 117, 12 (1960).

[60b] T. D. Lee and C. N. Yang.
Many-Body Problem in Quantum Statistical Mechanics. IV. Formulation in Terms of Average Occupation Number in Momentum Space.
Phys. Rev. 117, 22 (1960).

[60c] T. D. Lee and C. N. Yang.
Many-Body Problem in Quantum Statistical Mechanics. V. Degenerate Phase in Bose-Einstein Condensation.
Phys. Rev. 117, 897 (1960).

[60d] T. D. Lee and C. N. Yang.
Theoretical Discussions on Possible High-Energy Neutrino Experiments.
Phys. Rev. Lett. 4, 307 (1960).

[60e] T. D. Lee and C. N. Yang.
Implications of the Intermediate Boson Basis of the Weak Interactions: Existence of a Quartet of Intermediate Bosons and their Dual Isotopic Spin Transformations Properties.
Phys. Rev. 119, 1410 (1960).

[60f] C. N. Yang.
The Many-Body Problem.
Lectures given at Latin American School of Physics, Centro Brasileiro de Pesquisas Fisicas, Rio de Janeiro, June 27 – August 7, 1960. Lecture notes taken by M. Bauer, Y. Chou, F. Guerin, C. A. Heras and F. Medina. In Manografias de Fisica VI. Rio de Janeiro: Centro Brasileiro de Pesquisas Fisicas, 1960.

[60g] C. N. Yang.
Imperfect Bose System.
Physica 26, S49, (1960).

[60h] C. N. Yang.
Some Theoretical Implications of High-Energy Neutrino Experiments.
Lecture given at Berkeley Conference on High Energy Physics Experimentation, Sept. 12-14, 1960. Published in conference report, Univ. of Calif. Berkeley, 1960.

[61a] C. N. Yang.
Introductory Notes to the Article "Are Mesons Elementary Particles?".
Collect Papers of E. Fermi, Vol. 2, Chicago: Univ. of Chicago Press, (1965), p. 673.

[61b] T. D. Lee and C. N. Yang.
Some Considerations on Global Symmetry.
Phys. Rev. 122, 1954 (1961).

[61c] N. Byers and C. Yang.
Theoretical Considerations Concerning Quantized Magnetic Flux in Superconducting Cylinders.
Phys. Rev. Lett. 7, 46 (1961).

[61d] T. D. Lee, R. Serber, G. C. Wick, and C. N. Yang.
Some Theoretical Considerations on the Desirability of a 300 to 1000 BeV Proton Accelerator.
In Experimental Program Requirements for a 300-1000 BeV Accelerator, BNL 772 (T-290) p. 15 (1961).
Brookhaven National Lab. Aug. 1961 .

[61e] T. D. Lee, P. Markstein, and C. N. Yang.
Production Cross Section of Intermediate Bosons by Neutrinos in the Coulomb Field of Protons and Iron.
Phys. Rev. Lett. 7, 429 (1961).

[61f] C. N. Yang.
The Future of Physics.
Panel Discussion at the MIT Centennial Celebration, April 8, 1961. Unpublished.

物理學的前景
《楊振寧談科學發展》，張美曼編，
八方文化企業公司(1992)，p.15

[61g] F. Gursey and C. N. Yang.
S-State Capture in K^-p Atoms Colliding with H Atoms.
Written in May 1961. Unpublished.

[62a] C. N. Yang.
Symposium Discussion, November 4, 1961, Washington D. C. "Applied Mathematics: What is Needed in Research and Education?"
SIAM Review 4, 297 (1962).

[62b] S. B. Treiman and C. N. Yang.
Tests of the Single-Pion Exchange Model.
Phys. Rev. Lett. 8, 140 (1962).

[62c] C. N. Yang.
Elementary Particles, A Short History of Some Discoveries in Atomic Physics.
Princeton University Press (1962).

[62c.2] 基本粒子發現史
楊振玉譯, 上海科技出版社(1963).

[62c.3] 基本粒子
錢相、林多樑譯, 台灣中華書局 (1966).

[62c.4] ***Elementarpartiklar***
Translated by H.I. Gedin, Wahlström and Widstrand, Stockholm (1964).

[62c.5] ***Elementarteilchen***
Translated by I.M. Lambeck, Walter de Gruyter, Berlin (1971).

[62c.6] 素粒子之發現
林一譯, Misuzu Shobo (1968).

[62c.7] ***La scoperta della particella elementari***
Translated by A. Loinger, Paola Boringhieri, Torino (1964).

[62c.8] ***Elementary Particles***
Translated into Russian by A.M. Moiseeva, State Atomic Energy Publishing Bureau (1963).

[62c.9] 基本粒子及其相互作用
楊振玉、范世藩等譯, 湖南教育出版社 (1999).

[62d] T. D. Lee and C. N. Yang.
High Energy Neutrino Reactions Without Production of Intermediate Bosons.
Phys. Rev. 126, 2239 (1962).

[62e] M. E. Rose and C. N. Yang.
Eigenvalues and Eigenvectors of a Symmetric Matrix of 6j Symbols.
J. Math. Phys. 3, 106 (1962).

[62f] T. F. Hoang and C. N. Yang.
A Possible Method of Measuring the Fraction of $\Delta Q/\Delta S = -1$ Decay in the K_1 - K_2 Complex.
CERN Internal Report, 4010/TH. 276, May 28, 1962.

[62g] T. D. Lee and C. N. Yang.
Obituary for Dr. Shih-Tsun Ma.
Unpublished.

[62h] C. N. Yang.
Talk at CERN, July 7, 1962.
CERN preprint (1962).

[62i] T. D. Lee and C. N. Yang.
Theory of Charged Vector Mesons Interacting with the Electromagnetic Field.
Phys. Rev. 128, 885 (1962).

[62j] C. N. Yang.
Concept of Off-Diagonal Long-Range Order and the Quantum Phases of Liquid He and of Superconductors.
Rev. Mod. Phys. 34, 694 (1962).

[63a] C. N. Yang.
Mathematical Deductions from Some Rules ConcerningHigh-Energy Total Cross Sections.
J. Math. Phys. 4, 52 (1963).

[63b] C. N. Yang.
Some Properties of the Reduced Density Matrix.
J. Math. Phys. 4, 418 (1963).

[63c] C. N. Yang.
Remarks on Weak Interactions.
Proceedings of the Eastern Theoretical Physics Conference, ed. by M. E. Rose. New York: Gordon and Breach, 1963.

[63d] R. J. Oakes and C. N. Yang.
Meson-Baryon Resonances and the Mass Formula.
Phys. Rev. Lett. 11, 174 (1963).

[63e] C. N. Yang.
The Mass Formula of SU_3.
In Some Recent Advances in Basic Sciences, Vol. 1.
New York: Academic Press, 1966.

[64a] C. N. Yang.
Some Theoretical Considerations Concerning the Neutrino Experiments.
In Proc. of Weak Interaction Conference, Brookhaven National Lab, BNL 837 (C-39), p. 249, BNL, 1964.

[64b] C. N. Yang.
Computing Machines and High-Energy Physics.
Proceedings of the IBM Scientific Computing Symposium on Large Scale Problems in Physics, Dec. 1963, p. 65, IBM, 1964.

[64c] N. Byers and C. N. Yang.
Physical Regions in Invariant Variables for η Particles and the Phase-Space Volume Element.
Rev. Mod. Phys. 36, 595 (1964).

[64d] N. Byers and C. N. Yang.
Phenomenological Analysis of Reactions such as $K^- + p \rightarrow \Lambda + \omega$.
Phys. Rev. 135, B796 (1964).

[64e] C. N. Yang and C. P. Yang.
Critical Point in Liquid-Gas Transitions.
Phys. Rev. Lett. 13, 303 (1964).

[64f] Tai Tsun Wu and C. N. Yang.
Phenomenological Analysis of Violation of CP Invariance in Decay of K° and K°.
Phys. Rev. Lett. 13, 380 (1964).

[64g] C. N. Yang.
Round-Table Discussion on High-Energy Physics. APS Washington Meeting.
Phys. Today, 17, p. 50 (Nov. 1964).

[64h] F. Dyson, A. Pais, B. Stromgren, and C.N. Yang.
To J. Robert Oppenheimer on His Sixtieth Birthday.
Rev. Mod. Phys. 36, 507 (1964).

[65a] Tai Tsun Wu and C. N. Yang.
Some Speculations Concerning High-Energy Large Momentum Transfer Processes.
Phys. Rev. 137, B708 (1965).

[65b] C. N. Yang.
Some Considerations Concerning Very High Energy Experiments.
Nature of Matter: Purposes of High Energy Physics,
ed. by Luke C. L. Yuan, BNL 888 (T-360), P. 74, BNL, 1965.

[65c] C. N. Yang.
Report of the Theoretical Physics Panel to the Physics Survey Committee, Feb. 20, 1965.
Physics: Survey and Outlook, Reports on the Subfields of Physics, p. 159. NAS, NRC, (1966).

[65d] C. N. Yang.
Phenomenological Description of K Decay.
Proceedings of Int'l Conf. on Weak Interactions, p. 29, ANL 7130, 1965.

[65e] C. N. Yang.
Symmetry Principles in Physics.
Vistas in Research, Vol. 3, New York: Gordon and Breach, 1966. Also in Physics Teachers 5, p. 311, Oct. 1967.

[65f] N. Byers, S. W. MacDowell and C. N. Yang.
CP Violation in K Decay.
High Energy Physics and Elementary Particles, Vienna: International Atomic Energy Agency, Vienna 1965, p. 953.

[65g] C. N. Yang.
Statement at Public Hearing, Subcommittee on Research Development, Radiation, Congressional Record March 3, 1965.
Congressional Records.

[65h] C. N. Yang.
Speech on Last Day of Kyoto Conference, Commemorating the 30th Anniv. of Meson Theory, Sept. 30, 1965.
Unpublished.

[66a] C. N. Yang and C. P. Yang.
One-Dimensional Chain of Anisotropic Spin-Spin Interactions.
Phys. Lett. 20, 9 (1966); (Errata) Phys. Lett. 21, 719 (1966).

[66b] N. Byers and C. N. Yang.
πp ***Charge-Exchange Scattering and a "Coherent Droplet" Model of High Energy Exchange Processes.***
Phys. Rev. 142, 976 (1966).

[66c] C. N. Yang
Remarks at the Dedication of the Einstein Stamp, March 14, 1966.
Unpublished.

[66d] C. N. Yang and C. P. Yang.
Ground State Energy of a Heisenberg-Ising Lattice.
Phys. Rev. 147, 303 (1966).

[66e] C. N. Yang and C. P. Yang.
One-Dimensional Chain of Anisotropic Spin-Spin Interactions. I. Proof of Bethe's Hypothesis for Ground State in a Finite System.
Phys. Rev. 150, 321 (1966).

[66f] C. N. Yang and C. P. Yang.
One-Dimensional Chain of Anisotropic Spin-Spin Interactions. II. Properties of the Ground State Energy Per Lattice Site for an Infinite System.
Phys. Rev. 150, 327 (1966).

[66g] C. N. Yang and C. P. Yang.
One-Dimensional Chain of Anisotropic Spin-Spin Interactions. III. Applications.
Phys. Rev. 151, 258 (1966).

[66h] R. L. Mills and C. N. Yang.
Treatment of Overlapping Divergences in the Photon Self-Energy Function.
Progress of Theoretical Physics Supplement No. 37 and 38, 507 (1966).

[66i] C. N. Yang.
Summary of the Conference.
Proceedings of the Conference on High Energy Two-Body Reactions, Stony Brook, April 1966. Unpublished.

[66j] C. N. Yang.
Quantum Lattice Gas and the Heisenberg-Ising Anti-ferromagnetic Chain.
Proc. of the Eastern Theoretical Conf., p. 215, Providence, R.I., Brown University (1966).

[66k] M. Dresden, A. Lundby and C. N. Yang.
Proceedings of the Conference on High Energy Two-Body Reactions.
Stony Brook Conference, April 22-23, 1966 (unpublished).

[67a] F. Abbud, B. W. Lee, and C. N. Yang.
Comments on Measuring Re (A_2/A_0) in the decay $K_s^o \rightarrow \pi+\pi$.
Phys. Rev. Lett. 18, 980 (1967).

[67b] T. T. Chou and C. N. Yang.
Some Remarks Concerning High Energy Scattering.
High Energy Physics and Nuclear Structure, Amsterdam: North-Holland, 1967, p. 384.

[67c] B. Sutherland, C. N. Yang, and C. P. Yang.
Exact Solution of a Model of Two-Dimensional Ferroelectrics in an Arbitrary External Electric Field.
Phys. Rev. Lett. 19, 588 (1967).

[67d] T. T. Wu and C. N. Yang.
Some Solutions of the Classical Isotopic Gauge Field Equations.
Properties of Matter Under Unusual Conditions, ed. by H. Mark and S. Fernbach. New York: Wiley Interscience, p. 349 (1969).

[67e] C. N. Yang.
Some Exact Results for the Many-Body Problem in One-Dimension with Repulsive Delta Function Interaction.
Phys. Rev. Lett. 19, 1312 (1967).
Detailed proof later published by M. K. Fung in J. Math. Phys. 22, 2017 (1981)

[68a] C. N. Yang.
S Matrix for the One-Dimensional N-Body Problem with Repulsive or Attractive Delta Function Interaction.
Phys. Rev. 168, 1920 (1968).

[68b] T. T.Chou and C. N. Yang.
Model of Elastic High-Energy Scattering.
Phys. Rev. 170, 1591 (1968).

[68c] T. T. Chou and C. N. Yang.
Possible Existence of Kinks in High-Energy Elastic pp Scattering Cross Section.
Phys Rev. Lett. 20, 1213 (1968).

[68d] T. T. Chou and C. N. Yang.
Model of High-Energy Elastic Scattering and Diffractive Excitation Processes in Hadron-Hadron Collisions.
Phys. Rev. 175, 1832 (1968).

[68e] C. N. Yang.
General Review of Some Developments in High Energy Physics in Recent Years.
Paper delivered at First Latin-American Congress, Mexico City, July 1968.
Primer Congreso Latino Americano de Fisica, p. 27 (1968).

[68f] C. N. Yang.
The Status and Problems of High-Energy Physics Today.
Science, Vol. 161, No. 3836, July 5, 1968.

[69a] C. N. Yang and C. P. Yang.
Thermodynamics of a One-Dimensional System of Bosons with Repulsive Delta-Function Interaction.
J. Math. Phys. 10, 1115 (1969).

[69b] M. Goldhaber and C. N. Yang.
The K° –K° System in p-p Annihilation at Rest.
Evolution of Particle Physics, ed. by M. Conversi. N.Y.: Academic Press (1969) p. 171.

[69c] J. Benecke, T. T. Chou, C. N. Yang, and E. Yen.
Hypothesis of Limiting Fragmentation in High Energy Collisions.
Phys. Rev. 188, 2159 (1969).

[69d] T. T. Chou and C. N. Yang.
Extrapolation of Elastic Differential πp Cross Section to Very High Energies and the Pion Form Factor.
Phys. Rev. 188, 2469 (1969).

[69e] C. N. Yang.
Hypothesis of Limiting Fragmentation.
Proceedings of the Third International Conference on High Energy Collisions, New York: Gordon and Breach, (1969) p. 509.

[69f] C. N. Yang, J. A. Cole, J. Good, R. Hwa, and J. Lee-Franzini, eds.
High Energy Collisions.
Proceedings of the Third International Conference on High Energy Collisions, New York: Gordon and Breach, (1969).

[70a] C. N. Yang.
Charge Quantization, Compactness of the Gauge Group, and Flux Quantization.
Phys. Rev. D1, 2360 (1970).

[70b] C. N. Yang.
Some Exactly Soluble Problems in Statistical Mechanics. (Lectures given at the Karpacz Winter School of Physics, February 1970.)
Proceedings of the VII Winter School of Theoretical Physics in Karpacz, Univ. of Wroclaw, 1970.
Translated into Chinese in Cryogenics and Superconductivity 17, no. 3, p. 1 (1989).

[70c] T. T. Chou and C. N. Yang.
Remarks about the Hypothesis of Limiting Fragmentation.
Phys. Rev. Lett. 25, 1072 (1970).

[70d] C. N. Yang.
One-Dimensional Delta Function Interaction (Lecture given at the Battelle Institute Colloquium, (Sept. 1970).
Critical Phenomena in Alloys, Magnates, and Superconductors, ed. by R.E. Mills, E. Ascher, R. Jaffee, New York: McGraw Hill Book Co. (1971) p. 13.

[70e] C. N. Yang.
Symmetry Principles (In Physics).
Article for Encyclopedia Americana, 1970 edition, S.v. Physics.

[70f] C. N. Yang.
High-Energy Hadron-Hadron Collisions, Lecture given at the Kiev Conference, August 1970.
Proceedings of the Kiev Conference - Fundamental Problems of the Elementary Particle Theory, p. 131 (1970). Acad. of Sci. of the Ukranian SSR, 1970.

[70g] C. N. Yang.
Comments after Professor Brewer's Talk.
Proceedings of the VII Winter School of Theoretical Physics in Karpacz, 1970, Univ. of Wroclaw.

[73b] Alexander Wu Chao and Chen Ning Yang.
Opaqueness of pp Collisions from 30 to 1500 GeV/c.
Phys. Rev. D8, 2063 (1973).

[73c] Chen Ning Yang.
Geometrical Description of the Structure of the Hadrons. Lecture given at the International Symposium on High Energy Physics,Tokyo, July 23, 1973.
Proceedings of the International Symposium on High Energy Physics, Univ. of Tokyo, 1973, pp. 629-634.

[74a] Alexander Wu Chao and Chen Ning Yang.
Possible Relationship between the Ratio π^+/π^- and the Average Multiplicity.
Phys. Rev. D9, 2505 (1974).

[74b] Alexander Wu Chao and Chen Ning Yang.
Charge Correlation Between Two Pions in a Statistical Charge Distribution Among Hadrons.
Phys. Rev. D10, 2119 (1974).

[74c] Chen Ning Yang.
Integral Formalism for Gauge Fields.
Phys. Rev. Lett. 33, 445 (1974).

[74d] Chen Ning Yang.
Relationship between Correlation Function $\rho\eta$ and Fluctuation Phenomena.
Unpublished.

[75a] H. T. Nieh, Tai Tsun Wu, and Chen Ning Yang.
Possible Interactions of the J Particle.
Phys. Rev. Lett. 34, 49 (1975).

[75b] Tai Tsun Wu and Chen Ning Yang.
Some Remarks About Unquantized non-Abelian Gauge Fields.
Phys. Rev. D12, 3843 (1975).

[75c] Tai Tsun Wu and Chen Ning Yang.
Concept of Nonintegrable Phase Factors and Global Formulation of Gauge Fields.
Phys. Rev. D12, 3845 (1975).

[75d] Gu Chao-Hao and Yang Chen-Ning.
Some Problems on the Gauge Field Theories.
Scientia Sinica 18, 483 (1975).

[75e] Chen Ning Yang.
Gauge Fields.
Proceedings of Sixth Hawaiian Topical Conference in Particle Physics, Honolulu: University of Hawaii Press 1976, pp 487-561.

[75f] Chen Ning Yang.
Meccanica Statistica.
Twentieth Century Encyclopedia. Vol. 4, 1979, p. 53.

[76a] T. T. Chou and Chen Ning Yang.
Hadronic Matter Current Distribution Inside a Polarized Nucleus and a Polarized Hadron.
Nuclear Phys. B107, 1 (1976).

[76b] Tai Tsun Wu and Chen Ning Yang.
Static Sourceless Gauge Field.
Phys. Rev. D13, 3233 (1976).

[76c] Tai Tsun Wu and Chen Ning Yang.
Dirac Monopole Without Strings: Monopole Harmonics.
Nuclear Phys. B107, 365 (1976).

[76d] Tai Tsun Wu and Chen Ning Yang.
Dirac's Monopole Without Strings: Classical Lagrangian Theory.
Phys. Rev. D14, 437 (1976).

[76e] Chen Ning Yang.
Monopoles and Fiber Bundles.
Understanding the Fundamental Constituents of Matter,
ed. A. Zichichi. New York: Plenum, 1976, p. 53.

[76f] Chen Ning Yang.
Discussions "On Hadronic Current" with Dr. Leader and Others.
Understanding the Fundamental Constituents of Matter,
ed. A. Zichichi. New York: Plenum, 1976, p. 68.

[76g] Chen Ning Yang.
What Visits Mean to China's Scientists.
Reflections on Scholarly Exchanges with the People's Republic of China, ed. Anne Keatley (Committee on Scholarly Communication with the People's Republic of China) 1976.

[77a] M. L. Good, Y. Kazama, and Chen Ning Yang.
Possible Experiments to Study Incoherent-Multiple-Collision Effects in Hadron Production from Nuclei.
Phys. Rev. D15, 1920 (1977).

[77b] Tai Tsun Wu and Chen Ning Yang.
Some Properties of Monopole Harmonics.
Phys. Rev. D16, 1018 (1977).

[77c] Y. Kazama, Chen Ning Yang, and A. S. Goldhaber.
Scattering of a Dirac Particle with Charge Ze by a Fixed Magnetic Monopole.
Phys. Rev. D15, 2287 (1977).

[77d] Yoichi Kazama and Chen Ning Yang.
Existence of Bound States for a Charged Spin 1/2 Particle with an Extra Magnetic Moment in the Field of a Fixed Magnetic Monopole.
Phys. Rev. D15, 2300 (1977).

[77e] Chen Ning Yang.
Magnetic Monopoles, Fiber Bundles, and Gauge Fields.
Annals of N. Y. Academy of Science, 294, 86 (1977).

磁單極、纖維叢和規範場
《自然雜誌》第2卷第1期 (1979), p.10.

[77f] Chen Ning Yang.
Conformal Mapping of Gauge Fields.
Phys. Rev. D16, 330 (1977).

[77g] Chen Ning Yang.
Condition of Self-Duality for SU_2 Gauge Fields on Euclidean 4-Dimensional Space.
Phys. Rev. Lett. 38, 1377 (1977).

[77h] Gu Chao-Hao and Yang Chen-ning.
Some Problems on the Gauge Field Theories, II.
Scientia Sinica 20, 47 (1977).

[77i] Gu Chao-Hao and Yang Chen-ning.
Some Problems on the Gauge Field Theories, III.
Scientia Sinica 20, 177 (1977).

[77j] Chen Ning Yang.
Speech at the Benjamin W. Lee Memorial Session.
Unification of Elementary Forces and Gauge Theories, eds.
David B. Cline and Frederick E. Mills, p. xiii.
Harwood Academic Publishers, 1977.

[77k] Chen Ning Yang.
Symmetries in Physics.
Unification of Elementary Forces and Gauge Theories, eds.
David B. Cline and Frederick E. Mills, p. 3,
Harwood Acadmic Publishers, 1977.

[78a] Chen Ning Yang.
Generalization of Dirac's Monopole to SU_2 Gauge Fields.
J. Math. Phys. 19, 320 (1978).

[78b] T. T. Chou and Chen Ning Yang.
Possible Existence of a Second Minimum in Elastic pp Scattering.
Phys. Rev. D17, 1889 (1978).

[78c] Alexander W. Chao, Tai Tsun Wu, and Chen Ning Yang.
Some Inequalities in the Eikonal Approximation.
Ta-You Wu Festschrift: Science of Matter, ed. S. Fujita.
New York: GordonBreach, 1978.

[78d] Chen Ning Yang.
Interaction of a Static Magnetic Monopole with a Dirac Positron.
Proceedings of INS Intern'l Symp. on New Particles and the Structure of Hadrons, ed. by K. Fujikawa, Y. Hara, T. Maskawa, H. Terazawa, Inst. for Nuclear Study, Univ. of Tokyo, 1977.

[78e] Tu Tung-sheng, Wu Tai-Tsun, and Yang Chen-ning.
Interaction of Electrons, Magnetic Monopoles, and Photons (I).
Scientia Sinica 21, 317 (1978).

[78f] Ling-Lie Wang and Chen Ning Yang.
Classification of SU_2 Gauge Fields.
Phys. Rev. D17, 2687 (1978).

[78g] Chen Ning Yang.
SU_2 Monopole Harmonics.
J. Math. Phys. 19, 2622 (1978).

[78h] Chen Ning Yang.
Developments in the Theory of Magnetic Monopoles.
Proceedings of the 19th Int'l Conf. on High Energy Physics, Tokyo 1978, eds. S. Homma, M. Kawaguchi, and H. Miyazawa, p. 497. Phys. Soc. of Japan 1978.

[78i] Gu Chao-hao, Hu He-sheng, Shen Chun-li, and Yang Chen-ning.
A Geometrical Interpretation of Instanton Solutions in Euclidean Space.
Scientia Sinica 21, 767 (1978).

[78j] Gu Chao-hao, Hu He-sheng, Li Da-qian, Shen Chun-li, Xin Yuan-long, and Yang Chen-ning.
Riemannian Spaces with Local Duality and Gravitational Instantons.
Scientia Sinica 21, 475 (1978).

[78k] Chen Ning Yang.
Pointwise SO_4 Symmetry of the BPST Pseudoparticle Solution.
University Studies 66, 243 (1980).

[79a] T. T. Chou and Chen Ning Yang.
Elastic Hadron-Hadron Scattering at Ultrahigh Energies and Existence of Many Dips.
Phys. Rev. D19, 3268 (1979).

[79b] Max Dresden and Chen Ning Yang.
Phase Shift in a Rotating Neutron or Optical Interferometer.
Phys. Rev. D20, 1846 (1979).

[79c] Chen Ning Yang.
Fiber Bundles and the Physics of the Magnetic Monopole.
The Chern Symposium 1979, eds. W. Y. Hsiang, et. al., p. 247. (Springer Verlag, 1980).

[79d] Chen Ning Yang.
Panel Discussion.
Some Strangeness in the Proportion, ed. H. Woolf. Reading, Mass.: Addison-Wesley, 1980, p. 500.

[79e] Chen Ning Yang.
Geometry and Physics.
To Fulfill a Vision, ed. Y. Ne'eman. Reading, Mass.: Addison-Wesley, 1981, pp. 3-11.

幾何與物理
葛墨林、羅春榮譯 (1982)

[80a] T. T. Chou and Chen Ning Yang.
Geometrical Model of Hadron Collisions.
Proc. of the 1980 Guangzhou Conf. on Theo. Particle Physics,
Beijing: Sci. Press, 1980, p. 317.

[80b] Chen Ning Yang.
Einstein and the Physics of the Second Half of the Twentieth Century.
Proceedings of the Second Marcel Grossmann Meeting on General Relativity, ed. R. Ruffini, p. 7, North Holland, 1982.
Also appeared in slightly changed version, without the 2 leaves diagram in
Einstein's Impact on Theoretical Physics.
Physics Today 33, 42 (1980).
Compare comments on R. Herman's letter to the Editor,
Physics Today 33, 11 (1980).

愛因斯坦對理論物理的影響
《自然辯証法通訊》 No. 2 (1981)

愛因斯坦和二十世紀後半期的物理學
《物理教學》 第一期 (1980)

[80c] T.T. Chou and Chen Ning Yang.
Dip and Kink Structures in Hadron-Nucleus and Hadron-Hadron Diffraction Dissociation.
Phys. Rev. D22, 610 (1980).

[80d] Chen Ning Yang and Chen Ping Yang.
Does Violation of Microscopic Time Reversal Invariance Lead to the Possibility of Entropy Decrease?
Trans. N. Y. Academy of Sciences 40, 267 (1980).

[81a] T. T. Chou and Chen Ning Yang.
Dip Movement in pp and pp Elastic Collisions.
Phys. Rev. Lett. 46, 764 (1981).

[81b] Chen Ning Yang.
AGS 20th Anniversary Celebration.
BNL, 1980, p. 85.

[81c] T. T. Chou, T. T. Wu, and C. N. Yang.
Asymptotic Behavior at Very High Energies for Hadron-Hadron Collisions.
Phys. Rev. D26, 328 (1982).

[81d] C. N. Yang.
Lectures on Frontiers in Physics.
(The lst B.W. Lee Memorial Lecture on Physics, January 29 - February l, 1980, Seoul, Korea. Notes by K. S. Soh, J. H. Yee, I. K. Koh and S. K. Han), Korean Physical Society & Korean Science and Engineering Foundation, 1981.

[82a] Chen Ning Yang.
Bound States for e-g System.
in Monopoles in Quantum Field Theory, eds. N. S. Craigie, P. Goddard, and W. Nahm, p. 237, World Scientific 1982.

[82b] Chen Ning Yang.
Bound States for the e-g System (in Chinese).
in Festschrift for Prof. Chou Pei-yuan, Essays on Theoretical Physics and Mechanics, eds. J. S. Wang et. al., p. 7, Science Publications 1982.

電子與磁單極的束縛態

[82c] Chen Ning Yang.
Joseph Mayer and Statistical Mechanics.
International Journal of Quantum Chemistry: Quantum Chemistry Symposium 16, pp. 21-24 (1982), John Wiley & Sons, Inc..

[82d] T. T. Chou and Chen Ning Yang.
Remarks on Multiplicity Fluctuations and KNO Scaling in pp Collider Experiments.
Phys. Lett. 116B, 301 (1982).

[82e] Chen Ning Yang.
Flux Quantization, A Personal Reminiscence.
in Near Zero New Frontiers of Physics pp. 252-256 (1988), ed. by J. D. Fairbank, B. S. Deaver, Jr., C. W. F. Everitt, P. F. Michelson, W. H. Freeman and Company

[82f] C. N. Yang.
Why Research.
Speech given at the Research Award Dinner at the University of Georgia, April 1982.

[82g] Chen Ning Yang.
The Discrete Symmetries P, T and C.
Journal de Physique, Colloque C8, Supplement au No. 12 December 1982 p. C8-439.

[82h] Chen Ning Yang.
Beauty and Theoretical Physics.
In The Aesthetic Dimension of Science, Ed. Deane W. Curtin, pp. 25-40, and 107-145 Philosophical Library N.Y. 1982.

[83a] Chen Ning Yang.
The Spin.
Published in AIP Conference Proceedings #95 High Energy Spin Physics - 1982, p. 1, ed. Gerry M. Bunce (1983).

[83b] Yang Chen-ning.
Symmetry and the Physics of the 20th Century.
United Bulletin 1981-1982, Chinese University of Hong Kong.

對稱與二十世紀的物理
香港中文大學聯合校刊 1981-1982

[83c] T. T. Chou and Chen Ning Yang.
Elastic Scattering at CERN Collider Energy and the Geometrical Picture.
Phys. Lett. 128B, 457 (1983).

[83d] Chen Ning Yang.
Magnetic Monopoles and Fibre Bundles.
Accademia Nazionale Dei Lincei 57, Aspetti Matematici delle Teoria della Relativita 5-6 giugno 1980, p. 5 (1983).

[83e] Chen Ning Yang.
Gauge Field Lectures I, II, III.
Published in Gauge Interactions Theory and Experiments, ed. Antonino Zichichi, pp. 1-64 Plenum Press 1984.

[83f] Chen Ning Yang.
Forty Years as Student and Teacher.
Chinese University Bulletin Supplement 5,
Twentieth Anniversary Lecture,
Chinese University of Hong Kong.

讀書教學四十年
香港中文大學校刊附刊五
香港中文大學二十週年紀念講座

Chen Ning Yang.
Forty Years of Study and Teaching.
Translated into English by Jiang Ming-shan.

[83g] Chen Ning Yang.
Gauge Fields, Electromagnetism and the Bohm-Aharonov Effect.
Proc. Int'l Symp. Foundations of Quantum Mechanics, Tokyo, 1983, edited by S. Kamefuchi et. al., Physical Society of Japan, Tokyo (1984) pp. 5-9.

[83h] Chen Ning Yang.
Fields and Symmetries - Fundamental Concepts in 20th Century Physics.
Gakushuin University Pub. #2, p. 27 (1984).

[85d] Chen Ning Yang.
Particle Physics in the Early 1950s.
In Pions to Quarks, eds. Laurie M. Brown, Max Dresden, Lillian Hoddeson, p. 40, Cambridge Univ. Press (1989).

[85e] T. T. Chou and Chen Ning Yang.
Do Two Jet Events in e^+e^- Collision Observe KNO Scaling?
Proceedings of Kyoto Int'l Symp. The Jubilee of the Meson Theory, Kyoto, Japan, August 15-17, 1985.

[85f] Chen Ning Yang.
Remarks on the 1935 Paper of Yukawa.
Progress of Theoretical Physics, Supplement #85, pp. 11-12 (1985).

對湯川秀樹 1935 年的文章的評價
《自然雜誌》9 卷 11 期 (1986)，李炳安、張美曼譯

[85g] Chen Ning Yang and T. T. Chou.
Spin of Electrons, Hadrons and Nuclei.
Proceedings of Sixth Int'l Symp. Polarized Phenomena in Nuclear Physics, Osaka, 1985, J. Physical Society of Japan 55 (1986), Supplement pp. 53-57.

[85h] Chen Ning Yang.
Is KNO Scaling Valid for Two-Jet Events in e^+e^- Collisions?
Particles and detectors, Festschrift for Jack Steinberger, eds. K. Kleinknecht and T. D. Lee, p. 279 (1986), Springer-Verlag.

[85i] T. T. Chou and Chen Ning Yang.
A Unified Physical Picture: Narrow Poisson-like Distribution for e^+e^- Two-Jet Events and Wide Approximate KNO Distribution for Hadron-Hadron Collisions.
Physics Lett. 167B, 453 (1986). Erratum: Physics Lett. 171B, 486 (1986).

[85j] Chen Ning Yang.
Hermann Weyl's Contribution to Physics.
A Hermann Weyl 1885 Centenary Lecture, ETH, 1985
Hermann Weyl 1885-1985, ed. K. Chandrasekharan, p. 7,
Springer-Verlag l986.

魏爾對物理學的貢獻
《自然雜誌》9 卷 11 期 (1986) ，李炳安、張美曼譯

[85k] Chen Ning Yang.
Forty Years as a Student and a Teacher.
Book, (In Chinese) Joint Publishing Company, Hong Kong.

《讀書教學四十年》
香港三聯書店 (1985)
[85k.2] 台灣盜印 (1988)
[85k.3] 檳榔出版社再版
[85k.4] 曉園（台灣）

[86a] Chen Ning Yang.
Symmetry and Modern Physics.
Contribution to Festschrift for Arthur M. Sackler, Feb. 1986.

[86b] T. T. Chou and Chen Ning Yang.
Should there be KNO Scaling for e^+e^- two-jet Events?
Lecture given at the second Asia Pacific Physics Conference
Bangalor, Jan. 1986. Int'l J. of Mod. Phys. A1, 415-420 (1986).

[86c] Chen Ning Yang.
Square root of minus one, complex phases and Erwin Schrodinger.
In Schrodinger Centenary Celebration of a Polymath, Ed.
C. W. Kilmister, p. 53, Cambridge University Press, 1987.

負一的平方根、複相位與薛定諤
《楊振寧演講集》, 寧平治、唐賢民、張慶華主編,
南開大學出版社 (1989), p. 516

[86d] Chen Ning Yang.
Complex Phases in Quantum Mechanics.
Lecture given at the ISQM Conference, Tokyo, September 1986.
Proceedings 2nd Int. Symp. Foundations of Quantum Mechanics, Tokyo, 1986, pp. 181-184.

[86e] Chen Ning Yang.
Influence of Box Size on Ground State Energy of Two Interacting Particles.
Contribution to the Festschrift for Prof. Ta-You Wu on his eightieth birthday. Chinese Journal of Physics, 25, 80 (1987).

[86f] R.de Bruyn Ouboter and Chen Ning Yang.
Thermodynamic Properties of Liquid ^{3}He-^{4}He Mixtures between 0 and 20 Atmosphere in the Limit of Absolute Zero Temperature.
Physica 144B, 127 (1987).

[86g] B.A. Li and Chen Ning Yang.
K. C. Wang and the Neutrino.
(in Chinese) Journal of Dialectics of Nature, No. 5, p. 34 (1986).

王淦昌先生與中微子
《自然辯証法通訊》5期，p.34 (1986)
《楊振寧談科學發展》, 張美曼編, 八方文化企業公司 (1992), p .301

[86h] Chen Ning Yang.
Some Reminiscences.
Mathematical Medley 14, 51-74 (1986)

[87a] Chen Ning Yang.
Generalization of Sturm-Liouville Theory to a System of Ordinary Differential Equations with Dirac Type Spectrum.
Commun. Math. Phys. 112, 205 (1987).

[87b] Bang An Li and C. N. Yang.
C. Y Chao, Pair Creation and Pair Annihilation.
Intern'l Journal of Math.Physics A4 1989, 4325.
Chinese translation in Nature Magazine (Shanghai) 13, 275-279 (1990)

趙忠堯與電子對產生和電子對湮滅
《自然雜誌》13 卷 5 期 ，p. 275-279 (1990)
《楊振寧談科學發展》, 張美曼編 , 八方文化企業公司 (1992), p. 315

[87c] T. T. Chou and Chen Ning Yang.
Experimental Support for a Unified Physical Picture: Narrow Multiplicity Distribution for e^+e^- and Wide Multiplicity Distribution for pp Collisions.
Phys. Letters B193, 531 (1987).

[87d] T. T. Chou and Chen Ning Yang.
A Unified Physical Picture of Multiparticle Emission: Wide Multiplicity Distribution for pp and Narrow Multiplicity distribution for e^+e^- Collisions.
Intern'l Journal of Modern Physics A2, 1727 (1987).

[87e] Chen Ning Yang.
Review of the Collected Papers of Albert Einstein, Vol. 1, The Early Years: 1879-1902.
Nature 330, Issue No. 6143, 30 (1987).

[87f] Chen Ning Yang.
Science, Education and Chinese Modernization Program.
Book, (In Chinese) People's Daily Publishing Co..

《科學、教育與中國現代化》
人民日報出版社

[87g] Chen Ning Yang.
A True Story.
Physics, Vol. 3, No. 16, p. 146 (in Chinese).

一個真的故事
《物理》16 卷 3 期，p. 146 (1987)
《楊振寧演講集》, 寧平治、唐賢民、張慶華主編,
南開大學出版社 (1989), p. 192

[87h] 漢城二日印象記
《明報月刊》1987年10月號, p. 72

[88a] Chen Ning Yang.
Science, Technology and Economic Development.

[88b] C. H. Gu and Chen Ning Yang.
A One Dimensional N Fermion Problem with Factorized S Matrix.
Comm. Math. Phys. 122, 105 (1989).

[88c] T. T. Chou and Chen Ning Yang.
Remarks on Multiplicity and Momentum Distributions in e^+e^- Collisions at W = 14, 34, 55, 100 and 120 GeV.
Phys. Ltrs. B212, 105 (1988).

[88d] Chen Ning Yang.
Journey Through Statistical Mechanics.
Int. J. Mod. Phys. B2, 1325 (1988).

在統計力學領域中的歷程
《楊振寧談科學發展》, 張美曼編, 八方文化企業公司 (1992), p.329

[88e] Y. W. Chan, A.F. Leung, Chen Ning Yang and K. Young, eds.
Third Asia Pacific Physics Conference, Volumes 1 and 2.
World Scientific Publishing.

[88f] Chen Ning Yang.
Selected Interview and Essays.
Edited by K. K. Phua and C. Y. Han. (In Chinese)
World Scientific Publishing.
Also published (in Chinese) as Ning Chuo Wu Qiao.

《寧拙毋巧 — 楊振寧訪談錄》 **(B88a)**
潘國駒、韓川元編 ，世界科技出版社 (1988)及明報出版社 (1988)

[89a] Shao-Jing Dong and Chen Ning Yang.
Bound States Between Two Particles In A Two or Three Dimensional Infinite Lattice With Attractive Kronecker δ-Function Interaction.
Rev. Math. Phys. 1, 139 (1989).

[89b] C. N. Yang and M. L. Ge, eds.
Braid Theory, Knot Theory and Statistical Mechanics.
World Scientific Publishing Company.

[89c] Chen Ning Yang.
Modern Physics and Warm Friendship.
In Lattice Dynamics and Semiconductor Physics, edited by J. B. Xia et al., World Scientific, 1990.
Chinese translation in Ming Pao Monthly, August 1991 (Hong Kong)

現代物理和熱情的友誼
《明報月刊》1991年8月號

現代物理與深厚情誼 — 我的朋友黃昆
《自然雜誌》, 14卷10期 (1991), p. 786

[89d] Chen Ning Yang.
η-pairing and ODLRO in a Hubbard Model.
Phys. Rev. Lett. 63, 2144 (1989).

[89e] Chen Ning Yang.
Interview by Bill Moyers in
Bill Moyers, ***A World of Ideas***, Doubleday, (1989).

[89f] Chen Ning Yang.
Maxwell's Equations, Vector Potential and Connections on Fiber Bundles.
Proceedings of the Gibbs Symposium, Yale University, May 15-17, 1989, editors: G.D. Mostow, D. G. Caldi, p. 1, American Mathematical Society pp. 253-254 (1989).
Proceedings 3rd Int. Symp. Foundations of Quantum Mechanics, Tokyo, 1989, pp. 383-384.

[89g] Chen Ning Yang.
Selected Lectures.
Book, (in Chinese) Nankai University Press, 1989.

《楊振寧演講集》 **(B89b)**
南開大學出版社 (1989)

[89h] Chen Ning Yang.
Remembering Mr. T. S. Tsen.
(in Chinese) Memorial Volume for T. S. Tsen, Jiang-su Educational Publisher, 1989.

回憶鄭桐蓀先生
《鄭桐蓀先生紀念冊》，江蘇教育出版社 (1989)

[90a] T. T. Chou and Chen Ning Yang.
Elastic Proton-Antiproton Scattering from ISR to Tevatron Energies.
Physics Letters B 244, 113 (1990).

[90b] Chen Ning Yang and S. C. Zhang.
SO_4 Symmetry in a Hubbard Model.
Modern Physics Letters B4, 977-984 (1990).

[90c] Chen Ning Yang.
From the Bethe-Hulthen Hypothesis to the Yang-Baxter Equation.
The Oskar Klein Memorial Lectures, Vol. 1, Ed. Gosta Ekspong, World Scientific 1991.

[71a] C. K. Lai and C. N. Yang.
Ground State Energy of a Mixture of Fermions and Bosons in One-Dimension with a Repulsive δ-Function Interaction.
Phys. Rev. A3, 393 (1971).

[71b] C. N. Yang.
Introductory Note on Phase Transitions and Critical Phenomena.
In Phase Transitions and Critical Phenomena, Vol. 1, ed. by C. Domb and M. S. Green, New York: Academic Press, 1971, p. 1.

[71c] T. T. Chou and Chen Ning Yang.
Hadron Momentum Distribution in Deeply Inelastic ep Collisions.
Phys. Rev. D4, 2005 (1971).

[72a] Chen Ning Yang.
Some Speculations on Colliding Beams of 100-GeV Protons and 15-GeV Electron-100-GeV Protons.
Isabelle Physics Prospects, BNL 17522. BNL,1972.

[72b] C. Quigg, Jiunn-Ming Wang, and Chen Ning Yang.
Multiplicity Fluctuation and Multiparticle Distribution Functions in High-Energy Collisions.
Phys. Rev. Lett. 28, 1290 (1972).

[72c] Chen Ning Yang.
Some Concepts in Current Elementary Particle Physics. Lecture given at the Trieste Conference in honor of P.A.M. Dirac, September 1972.
The Physicist's Conception of Nature, ed. J. Mehra, Dordrecht: D. Reidel (1972) pp. 447-453.

[73a] T. T. Chou and Chen Ning Yang.
Charge Transfer in High-Energy Fragmentation.
Phys. Rev. D7, 1425 (1973).

[90d] Chen Ning Yang.
Physics of the Twentieth Century.
Published in Chinese in Twenty First Century Vol. 1, No. 1, Oct. 1990, p. 67

二十世紀的物理學
《二十一世紀》第一期 (1990)

[90e] Chen Ning Yang.
Symmetry and Physics.
The Oskar Klein Memorial Lectures, Vol. 1, Ed. Gosta Ekspong, World Scientific 1991.
Chinese Translation in Twenty First Century, August 1991, p. 6 (Hong Kong).

對稱和物理學
《二十一世紀》第六期 (1991)

[90f] Chen Ning Yang.
Banquet Speech.
Proceedings of the 25th International Conference on High Energy Physics, p. ix, Eds. K. K. Phua and Y. Yamaguchi (South East Asia Theoretical Physics Association, 1991)

[90g] Chen Ning Yang.
The Formation of the Association of Asia Pacific Physical Societies is the Right Event at the Right Time.
Speech on August 13, 1990 at the Fourth Asia Pacific Physics Conference at Yonsei University, Seoul, Korea.
Published in AAPPS Newsletter Vol. 1, No. 3, p. 5 (1990)

[90h] T. T. Chou and C. N. Yang.
Experimental Support for the Geometrical Picture of Multiparticle Emission in e^+e^- Collisions.
Proceedings of the 25th International Conference on High Energy Physics, p. 1013, Eds. K. K. Phua and Y. Yamaguchi (South East Asia Theoretical Physics Association, 1991).

[90i] Chen Ning Yang.
Remembering Mr. Y. H. Woo.
(in Chinese) Y.H. Woo, Literature and History Publisher of China, 1990.

回憶吳有訓先生
《吳有訓》，中國文史出版社 (1990)

[91a] Chen Ning Yang.
SO(4) Symmetry in a Hubbard Model.
in Springer Proceedings in Physics, Volume 57 Evolutionary Trends in the Physical Sciences, Eds. M. Suzuki and R. Kubo, Springer-Verlag Berlin Heidelberg 1991.

[91b] Yuefan Deng and Chen Ning Yang.
Waring's Problem for Pyramidal Numbers.
Science in China (Series A) 37, 3 (1994).

[91c] Chen Ning Yang.
Comments after Prof. Wu's Talk.
Speech at the Honorary Degree Ceremony at the University of Michigan on May 3, 1991.
Published in OCPA Newsletter 1, No. 4, page 6 (1991).

[91d] Chen Ning Yang.
Remarks and Generalizations of SU_2 x SU_2 Symmetry of Hubbard Models.
Physics Letters 161 (1991) 292-294.

[91e] Chen Ning Yang.
S. S. Chern and I.
Chern - A Great Geometer of the Twentieth Century, Ed. by Shing-Tung Yau, p. 63, International Press Co. Ltd, Hong Kong, 1992.
Chinese Translation in Twenty First Century, No. 10, April 1992.

陳省身先生與我
《二十一世紀》第十期 (1992)

[91f] Chen Ning Yang.
To Jayme Tiomno.
in Frontier Physics eds. S. MacDowell, H. M. Nussenzveig, and R. A. Salmeron, World Scientific, 1991.

[92a] Yuefan Deng and Chen Ning Yang.
Eigenvalues and Eigenfunctions of the Hückel Hamiltonian for Carbon-60.
Physics Letters A170 (1992) 116-126.

[92b] Chen Ning Yang.
Remarks at the end of the Conference at the Advanced Research Laboratory, Hitachi, August 1992.
Quantum Control and Measurement, H. Ezawa et. al. eds. Elsevier, 1993.

[92c] Chen Ning Yang.
Reflections on the Development of Theoretical Physics.
Proceedings 4th International Symposium Foundation of Quantum Mechanics in the Light of New Technology, p. 3 (Japanese Journal of Applied Physics, Tokyo) 1993.
Published in AAPPS Bulletin Vol. 3, No. 3, p.12 (1993).
Chinese translation in World Science, No. 6 (1993).

關於理論物理發展的若干反思
《世界科學》第六期 (1993)

[92d] Chen Ning Yang.
Mr. K. L. Hiong.
(in Chinese) In Memory of Prof. King-Lai Hiong on the Occasion of His Birth of 100 Years,
Yunnan Educational Publishers, 1992.

熊慶來先生
《熊慶來紀念集》，雲南教育出版社 (1992)

[92e] Chen Ning Yang.
Mr. Zhou Pei-Yuan.
(in Chinese) in Distinguished Scientist Great Teacher, p. 104, Science and Technology Publisher of China, 1992.

周培源先生
《科學巨匠 師表流芳》，中國科學技術出版社 (1992)

[93a] Chen Ning Yang.
Fullerenes and Carbon 60.
Published in Salamfestschrif, A. Ali, J. Ellis,
S. Randjbar-Daemi, Eds.
World Scientific 1994.

[93b] Chen Ning Yang.
Deng Jia-Xian.
(In Chinese). Twenty-First Century, No. 17, June 1993, p. 56.
(In Chinese). People's Daily, August 21, 1994.
(In Chinese). Wen Hui Bao, August 23, 1994.
(In Chinese). Science, Vol. 46, No. 5, Sept. 1994.
(In Chinese). Wen Hui Du Shu Weekly News, March 4, 1995.

鄧稼先
《二十一世紀》第十七期 (1993)
《人民日報》, 1994年8月21日
《文匯報》, 1994年8月23日
《科學》, 第46卷第5期 (1994年9月)
《文匯讀書周報》, 1995年3月4日
《語文》第四冊, p.113, 人民教育出版社, 2001年11月
《語文》, p. 111, 高中一年級第一學期教材,
華東師範大學出版社, 2002年7月

[93c] Chen Ning Yang.
History of the Introduction of Modern Science into China.
(In Chinese). Ming Pao Monthly, No. 10, Oct. 1993, p. 12.
(In Chinese). Science, Vol. 47, No. 1, Jan. 1995.

近代科學進入中國的回顧與前瞻
《明報月刊》 1993年10月號, p. 12

近代科學進入中國的回顧與前瞻
《科學》第47卷第1期 (1995年1月)

[93d] T. T. Chou and Chen Ning Yang.
Eigenvalue Problem of the Hückel Hamiltonian for Carbon-240.
Physics Letters A183, 221 (1993).

[93e] Chen Ning Yang.
Remarks made at the Ceremony on November 11, 1993 at the American Philosophical Society.

[93f] Chen Ning Yang.
Foreword.
A Guide to Physics Problems, Part I, Mechanics, Relativity and Electrodynamics, Sidney B. Cahn and Boris E. Nadgorny, Plenum Press, New York, 1994.

[93g] C.N. Yang, M.L. Ge, and X.W. Zhou, eds.
Proceedings of the XXI International Conference on Differential Geometric Methods in Theoretical Physics.
World Scientific 1993.

[94a] T.T. Chou and Chen Ning Yang.
Test of Limiting Fragmentation in ep Collisions at DESY HERA.
Phys. Rev. D50, 590 (1994).

[94b] C.N. Yang, M.L. Ge, eds.
Braid Group, Knot Theory and Statistical Mechanics II.
World Scientific 1994.

[94c] C.N. Yang.
Mr. T.Y. Wu and Physics
(in Chinese). Proceedings of the symposium in celebration of President T.Y. Wu's retirement, Academia Sinica, Taiwan, July 1994.

吳大猷先生與物理
《吳大猷院長榮退學術研討會論文集》, 台灣中央研究院, 1994年7月

[94c.2] Chen Ning Yang.
Professor T. Y. Wu and Physics.
Chinese Journal of Physics, Vol. 35, No. 6-II, Taiwan, December 1997, p. 737.

[94d] C. N. Yang.
S.T. Yau: A World-class Mathematician.
(in Chinese). Ming Pao Monthly, No. 7, July 1994, p. 59.

丘成桐：世界級數學家
《明報月刊》 1994年7月號, p. 59

[94e] C.N. Yang.
Laws of Nature, Science, Technology, and Human Society.
Draft of talk given on June 9, 1994 at the celebration of the 1,200th anniversary of the founding of the city of Kyoto.

[94f] T. T. Chou and Chen-Ning Yang.
Hückel Spectrum for Large Carbon Fullerenes and Tubes.
Phys. Letters A192, 406 (1994).

[94g] C.N. Yang.
Obituary for Julian Schwinger.
(in Chinese). Twenty-First Century, No. 24, August 1994.

悼念施溫格
《二十一世紀》第二十四期, 1994年8月

[94h] C. N. Yang.
Hong Kong Qiu Shi Science and Technologies Foundation presented "Distinguished Scientists Awards".
(in Chinese). Da Gong Bao, August 26, 1994.
(in Chinese). Reference News, August 30, 1994.

香港求是科技基金會頒發「傑出科學家獎」
《大公報》, 1994年8月26日。
《參考消息》, 1994年8月30日。

[94i] C. N. Yang.
Introduction
Published in A Commemorative Album of Donation Projects by Mr. Run Run Shaw, State Education Commission, PRC, January 1995.

前言
《邵逸夫先生贈款項目專刊》, 中國國家教育委員會, 1995年1月

[95a] C. N. Yang.
Conceptual Beginnings of Various Symmetries in Twentieth Century Physics.
Chinese Journal of Physics, Vol. 32, p. 1437 (1994).

[95b] C. N. Yang.
Julian Schwinger.
In Julian Schwinger The Physicist, The Teacher, and the Man, edited by Y. Jack Ng, World Scientific, 1996.

[95c] C. N. Yang.
Gu Chao-Hao and I.
In Collection of Papers on Geometry, Analysis and Mathematical Physics: In Honor of Professor Gu Chaohao, Ed. T.T. Li, World Scientific, 1997.

[95d] Chen Ning Yang.
Path Crossings with L. Onsager.
In Collected Works of Lars *Onsager*, Ed. P.C. Hemmer, H. Holden and S. Kjelstrup Ratkje, World Scientific, 1996.

[95e] C. N. Yang.
Vector Potential, Gauge Field and Connection on a Fiber Bundle.
In Quantum Coherence and Decoherence, edited by K. Fujikawa and Y.A. *Ono*, Elsevier Science B.V., 1996.

[95f] C.N. Yang.
Again Ten Years.
Book, (in Chinese), China Times Publishing Co., 1995.

《讀書教學再十年》
時報出版社 (1995)

[95g] C. N. Yang.
Remarks About Some Developments in Statistical Mechanics.
Published in AAPPS Bulletin, Vol. 5, Nos. 3 & 4, p. 2 (1995).

[96a] T.T. Chou, C. N. Yang, and L.H. Yu.
Bose-Einstein Condensation of Atoms in a Trap.
Phys. Rev. A53, 4257 (1996).

[96b] T.T. Chou, Chen Ning Yang, and L.H. Yu.
Momentum Distribution for Bosons with Positive Scattering Length in a Trap.
Phys. Rev. A55, 1179 (1997).

[96c] Chen Ning Yang.
Bose Einstein Condensation in a Trap.
Int. J. Mod. Phys. B, 11, 683 (1997).

[97a] Chen Ning Yang.
Beauty and Physics.
(in Chinese). Twenty-First Century, No. 40, April 1997.
(in Chinese). Science Monthly, Vol. 28, No. 9, September 1997.

美與物理學
《二十一世紀》第四十期, 1997年4月
《科學月刊》第二十八卷第九期, 1997年9月
《物理》第三十一卷第四期, 2002年4月
《文明》2002年, Nos. 7 & 8, 總第八、九期

[97b] C. N. Yang.
Vector Potential, Gauge Field and Connection on a Fiber Bundle.
Tsinghua Science and Technology, Vol. 3, No. 1, p.861, Journal of Tsinghua University, March 1998.

[97c] Chen Ning Yang.
"***Redeeming Sense of Shame.***"
(in Chinese). Ming Pao, July 20, 1997.

從國恥講起
《明報》, 1997年7月20日。

[97d] T. T. Chou and Chen Ning Yang.
Exact Solution of the Vibration Problem for the Carbon-60 Molecule.
Physics Letters A235, 97 (1997).

[97e] Chen Ning Yang.
Preface.
Published in Vanished Springs, Memoir of Prof. XU Yuan-zhong, Chinese Literature Press, 1998.

[97f] Chen Ning Yang.
Father and I.
(In Chinese). Twenty First Century, No. 44, December 1997.

父親和我
《二十一世紀》, 第四十四期, 1997年12月

[97f.2] Chen Ning Yang.
Father and I.
(English translation.)

[97g] Chen Ning Yang.
Address on the occasion of receiving an Honorary Degree from CUHK, December 11, 1997.
Chinese University Bulletin, Spring-Summer 1998.

[98a] Chen Ning Yang.
Collected Papers of Chen Ning Yang.
Book, (in Chinese), Vols. 1 and 2, Hua-dong Normal University Publishing Co., Shanghai, 1998.

《楊振寧文集》
上、下冊, 上海華東師範大學出版社 (1998)

[98b] Peter B. Kahn, Chen Ning Yang, Martin L. Perl, and Helen R. Quinn.
Obituaries for Max Dresden.
In Physics Today, p. 90, June 1998.

[98c] Chen Ning Yang.
Sound Velocity in a Boson Condensate.
APCTP Bulletin, p. 3, October 1998.

[99a] Chen Ning Yang.
Reflections on Daniel Tsui's Winning the Nobel Prize.
(in Chinese). The Joy of the Search for Knowledge, A Tribute to Professor Dan Tsui, eds. C.Y. Wong, S.I. Lo, and S.Y. Lo, World Scientific, 1999.

從崔琦榮獲諾貝爾獎談起
《求知樂》敬賀崔琦教授的文集, 世界科學出版社, 1999.

[99b] Chen Ning Yang.
Quantization, Symmetry and Phase Factor — Thematic Melodies of 20th Century Theoretical Physics

[00a] Chen Ning Yang.
我們
《明報月刊》2001年1月號

[00b] Chen Ning Yang.
中國文化與科學
《新亞生活》2000年2月

[00c] Chen Ning Yang.
中國現代文學館與魯迅頭像
《光明日報》2000年9月21日

[00d] Chen Ning Yang.
寄語新世紀
《光明日報》2000年12月27日

[00e] Chen Ning Yang.
Introduction by Prof. C. N. Yang in T. Y. Wu Session
In OCPA 2000: Proceedings of the Third Joint Meeting of Chinese Physicists Worldwide, eds. by N. P. Chang, K. Young, H. M. Lai and C. Y. Wong, World Scientific Publishing Co., 2002.

[00f] Chen Ning Yang.
感懷恩師
《典範永存 — 吳大猷先生紀念文集》, 中華民國物理學會, 遠流出版事業股份有限公司, 2001年4月

[01a] Chen Ning Yang.
中興業 需人傑
《光明日報》2001年1月11日

[01b] Chen Ning Yang.
給北島

[01c] Chen Ning Yang.
Reading Wu Weishan's Authenticity, Purity and Simplicity.
(in Chinese). Published in Guangming Daily, May 9, 2001.
Published in The Rhyme of Charm, Sculpture and Painting of Wu Wei Shan, Hong Kong University of Science and Technology Center for the Arts, 2001.

讀吳爲山雕塑：真純樸
《光明日報》2001年5月9日
《神韻》— 吳爲山雕塑及繪畫展, 香港科技大學藝術中心出版, 2001年

[01d] Chen Ning Yang.
Inaugural Preface: Words from Prof. Chen Ning Yang.
Published in Annual Bulletin of NCTS, Physics Division, January 2001.

[01e] Chen Ning Yang.
Yangzhenning Kejiao Wenxuan.
Book, (in Chinese) Nankai University Press, 2001.

《楊振寧科教文選》
南開大學出版社 (2001)

[01f] Chen Ning Yang.
Enrico Fermi.
(in Chinese). Published in Guangming Daily, September 29, 2001.

他永遠腳踏實地 — 紀念恩芮科 • 費米誕生一百周年
《光明日報》, 2001年 9月29日

[01g] Chen Ning Yang.
Werner Heisenberg (1901-1976).

沃納 • 海森堡 **(1901-1976)**
《二十一世紀》第70期, 2002年 4月

[01h] Chen Ning Yang.
Science and Technology of the 21st Century.
Published in Bulletin for the 6th World Chinese Entrepreneurs Convention, 2001. (ISBN 7-80100-819-7).

二十一世紀的科技
《盛世華章, 第六屆世界華商大會專輯》, 中華工商聯會出版社, 2001. (ISBN 7-80100-819-7).

[01i] Chen Ning Yang.
八十自述
《情系中華》, 上海文史資料選輯,
上海市政協文史資料編輯部, 新華書店, 2001年1月

[01j] Chen Ning Yang.
當前世界科技發展趨勢和我們應採取的對策
《自然雜誌》24卷1期, p. 1

[02a] Chen Ning Yang.
「大學之道」導言
香港電台「大學之道」電視節目, 小冊子及VCD, 2002年

[02b] Chen Ning Yang.
Preface
(in Chinese). Published in Between Parting and Goodbye, edited by Yu Li-hua, Enlighten Noah Publishing, 2002

序
《在離去與道別之間》, 於梨華著, 瀛舟出版社, 2002年11月

[02c] Chen Ning Yang.
八旬杏滿清華園
《明報月刊》, 2002年8月號

The Speech at the Banquet on June 18, 2002.
Published in AAPPS Bulletin Vol. 12, No. 2, p. 30 (2002).

[02d] Chen Ning Yang.
楊振寧教授學術演講
《求是基金會2001》, 2002年7月

[02e] Chen Ning Yang.
Preface
(in Chinese). Published in Yang Chen Ning, edited by Yip Chung Man, Chinese University Press, 2002.

序
《人情物理楊振寧》, 葉中敏著, 中文大學出版社, 2002年

[02f] Chen Ning Yang.
Necessary Subtlety and Unnecessary Subtlety
Speech at Neutrinos and Implications for Physics Beyond the Standard Model, Stony Brook, October 11-13, 2002.
Modern Physics Letters A, Vol. 17, No. 34 (2002), p. 2229.

[02g] Chen Ning Yang.
Address by Professor Chen-ning Yang at HKUST, November 8, 2002.
Published in Tenth Congregation, Hong Kong University of Science and Technology, 2002.

楊振寧教授2002年11月8日在香港科技大學的演講
《香港科技大學第十屆榮譽學位頒授典禮》，香港科技大學，2002年.

看清歷史發展 努力創建人生
《明報月刊》，2003年1月號

[02h] Chen Ning Yang.
在熊秉明葬禮上的講話
《二十一世紀》第75期, 2003年 2月

(Revised 13 May, 2003)